D1030426

FOR CLIMATE'S SAKE!

A VISUAL READER OF CLIMATE CHANGE

LARS MÜLLER PUBLISHERS

Anyone who talks to friends, whether on the bus, in school, at work, or at social gatherings, will soon realize that when it comes to climate change, there are a lot of opinions but not much knowledge.

Access to this complex subject matter can be gained only if factual information is presented in a transparent and easy to understand manner, thus raising awareness and enabling the formation of an educated opinion. For this reason—and in view of the failure of the climate conference in Copenhagen in 2009—we decided to follow up on the book we published in 2006—*Who Owns the Water?*—and undertake a similar project, this time with the goal of providing a comprehensive view of the multifaceted nature of climate and climate change.

As in the previous book, numerous illustrations and striking photographic sequences enhance background information, reports, and case studies. We address key questions such as: How has climate developed since the Earth's beginnings? How does the climate system work? Do we know the true reasons for global warming? What measures and strategies do climate scientists suggest to avoid the worst? How can humans adapt to the inevitable consequences of climate change? Why do politicians have so much trouble agreeing on effective measures?

Together with a creative and committed team we undertook the challenge of dealing with this extremely complex matter in a way that would appeal to a wide audience while still being scientifically reliable. The thought-provoking images chosen by Claudia Klein and Nadine Unterharrer, and the knowledgeable texts by respected science journalists Heidi Blattmann and Martin Läubli, came together in a book aiming to appeal to its readers on an intellectual as well as emotional level.

Our scientific panel, crucial to the realization of this project, was a group of climatologists, mainly from the Swiss Federal Institute of Technology (ETH) Zurich, who made sure that this book, although presented in language for the lay reader, still held up to strict scientific standards. We owe a great deal to them all and would like to extend our thanks to Stefan Brönnimann, Andreas Fischlin, Nicolas Gruber, Gertrude Hirsch Hadorn, Volker Hoffmann, Reto Knutti, Christian Pohl, and Helmut Weissert.

A book project of this scale would not have been possible without the help of like-minded foundations and institutions that aim to make scientific knowledge accessible to wider audiences, conveying the importance of science in our society, especially in schools. We wish to express our thanks for the trust and generous support of the Mercator Foundation Switzerland, the Avina Foundation, the Hamasil Foundation, the G+B Schwyzer Foundation, and, not least, the Department of Environmental Sciences at ETH Zurich.

We all hope that this book, with its factual approach, will contribute to stimulating debate and encouraging the action we need to deal with our planet's most urgent problem.

The Editors

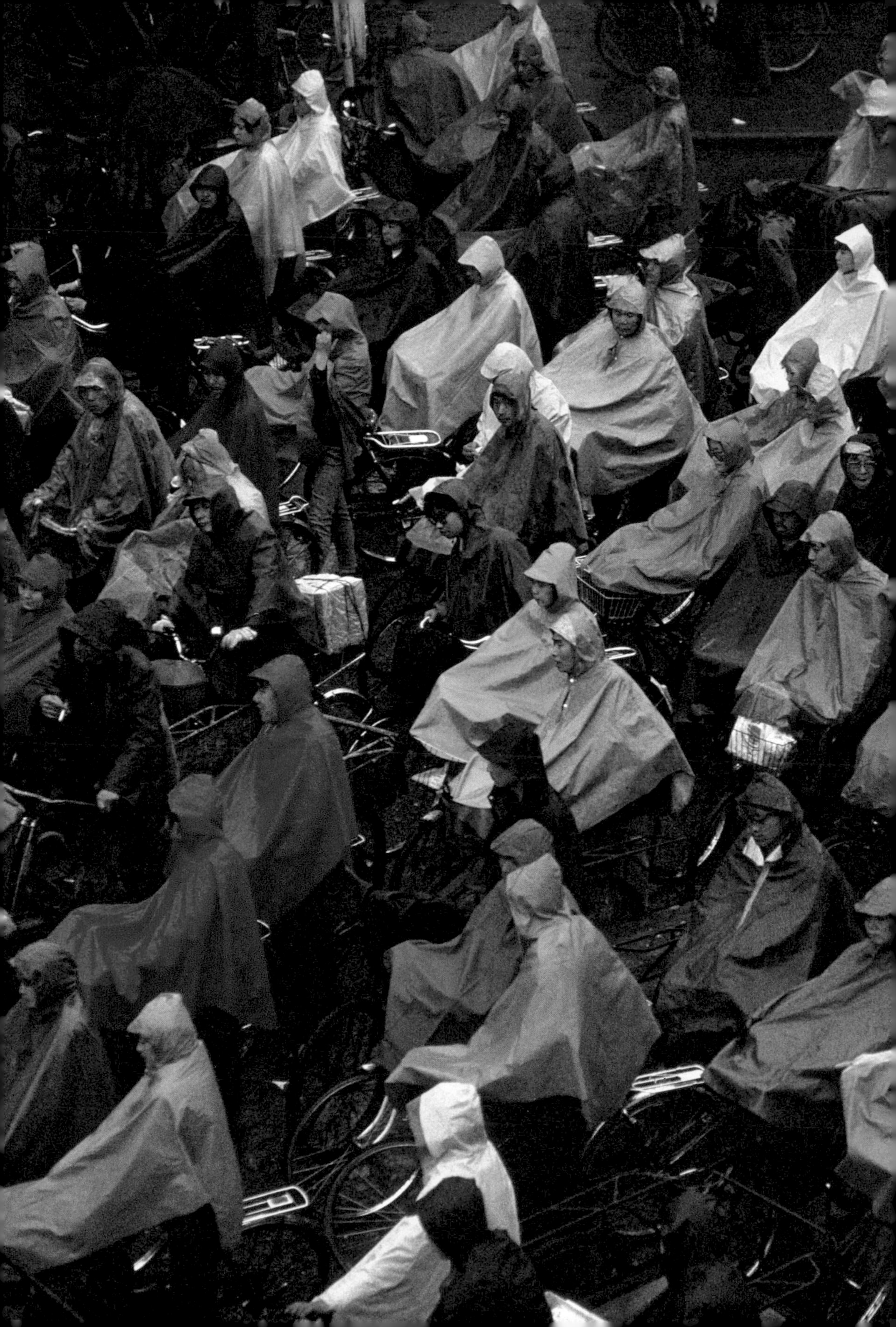

La pasta
לה פסטה
Restaurant מסעדה
16
OPEN
Heineken

escalier
אסקלייה
14

FOR CLIMATE'S SAKE!

THE CLIM
WON'T L
THE WOR

Christian Rentsch

MATE
EAVE
RLD COLD

There is really no such thing as climate. Everything we experience outdoors through our senses, everything we can feel and see is weather—encountered as heat or cold, observed as rain or snow.

In other words, climate is a concept invented by climatologists, an artificial construct that exists only in our minds. We communicate about climate using numbers, analytical data, statistics, descriptions of abstract processes, and computer models. We then make comparisons to add content to this construct—we compare the present with the past and we compare different regions to each other. We can find out how climate has changed only when we know about climate conditions in the past.

The reliability of scientific findings depends on the quality of the data available. Climatology is a young science—pioneers first attempted to understand climate events using astronomical and physical calculations only at the beginning of the nineteenth century. The thermometer has been in existence for just 350 years, and devices to measure carbon dioxide concentrations in the atmosphere were not invented until the last century. The technologies that enable climatologists to discover what Earth's climate was like thousands or even millions of years ago, by analyzing sediment layers, ice cores, glacier snow, and fossils, are even more recent. And only in the past few decades have supercomputers allowed us to calculate complex climatic interrelationships or simulate changes using computer models.

The climate system is highly complex, shaped by the intricate interaction of the sun, the atmosphere, the oceans, and continental landmasses. Other important "players" in the game are the Earth's orbit, greenhouse gases, types of landscape, vegetation, and, not least, humans. There is no doubt that early climatologists knew this, but they did not have the resources or the technology to collect data and manage computations on the climate system in all of its complexity. Climatology has become a comprehensive academic research area only since the middle of the twentieth century. Researchers from numerous disciplines are involved, whether they are physicists, chemists, biologists, geologists, oceanographers, meteorologists, or computer scientists—and even economists, historians, sociologists, and archeologists have entered this field.

It was this interdisciplinary cooperation that made it possible to piece together the individual parts of the puzzle that make up the climate system. There are still important pieces missing, and some interactions and feedback mechanisms in oceans or cloud formations, for instance, still haven't been explored enough. But by basing their work on numbers, analytical data, theoretical calculations, and computer models, climatologists today are able to come to very accurate conclusions on how the climate system works and how climate might develop under certain conditions.

Humans are the main element of uncertainty in attempts to predict future climatic trends. Today there is hardly any reasonable doubt that anthropogenic (man-made) greenhouse gas emissions are responsible for global warming. For the first time in Earth's history, humans can consciously decide on how global climate will develop. It is unprecedented in the history of humankind that decisions we make today will have such far-reaching consequences for future generations. Climate change will affect the entire human race for hundreds of years to come.

It is not easy to know what the effects of global warming will be. The rise in sea levels, changes in ocean currents and wind systems, the melting of mountain and inland glaciers in Antarctica and the Arctic are difficult to predict because they depend on very complex feedback mechanisms and long-range effects. They also depend on many specific regional and local conditions. Roughly every two years, the Intergovernmental Panel on Climate Change (IPCC) issues reports compiling the most recent scientific findings on climate change. However, the IPCC limits its commentary to relatively few, rather general statements in its concluding *Synthesis Report.*

It is even more difficult to forecast the real impact of climate change on people around the world. There is a huge difference between floods and famines hitting megacities with millions of inhabitants, or affecting rural, sparsely populated regions. The extent of damage depends largely on the means and resources available to afflicted populations to help them adapt to any expected changes in climate or protect themselves from unavoidable consequences. Climate change will have much graver impact on the poorer developing and emerging countries in tropical regions than on the northern industrial nations. The global community will need to confront these issues and face moral dilemmas.

Climatologists believe that we may still be able to mitigate global warming enough to limit irreversible damage and the ensuing complications that are now so hard to predict. But sweeping measures would have to be introduced over the next few decades for this to happen. First and foremost, we need to drastically reduce our consumption of coal, oil, and natural gas, and replace these fossil fuels with renewable energy sources. At the same time, the world must learn to conserve available energy and use it more efficiently to make sure that even the consumption of renewable energy does not increase uncontrollably.

But this will not be enough on its own. The plant and animal worlds, our planet's ecosystems, are not as adaptable as humans are. Oceans and the ecosystems absorb a large share of anthropogenic CO_2 emissions, but their ability to do so is limited. Should they lose this capacity, the pace of global warming will accelerate. A growing number of climatologists believe that we must adopt climate measures going far beyond energy issues and include a careful and sustainable approach to our natural environment.

The climate knows no national borders, which is why goals for mitigating climate change must be agreed upon at the international level. Climatologists, experts and politicians have been discussing this issue for nearly forty years. The United Nations Framework Convention on Climate Change, which has already been in existence for twenty years, does at least define the general goals of climate protection. With the signing of the 1997 Kyoto Protocol, the industrial nations—with the exception of the United States—committed themselves to reducing greenhouse gas emissions by at least 5 percent compared to 1990 levels during the first commitment period from 2008 to 2012. But another fourteen years went by before nations participating at the climate conference in Cancún could agree to set a two-degree Celsius (3.6°F) limit on the increase in global temperature.

Whether these targets will be reached, or whether industrial nations, together with emerging and developing countries, will succeed in the next few years in reaching a binding agreement on the measures and mechanisms needed to mitigate climate change, is quite uncertain. International climate policy has long been a pawn in the tug of war between conflicting political interests staked by the world's large powers.

“We are all passengers on the Titanic, although some of us are traveling first class.”

 Susan George, political scientist

pp. 2–3 London, United Kingdom. *Daniel Berehulak/Getty Images*

pp. 8–9 Seoul, South Korea. *Thomas Hoepker/Magnum Photos*
pp. 10–11 Shanghai, China. *Stuart Franklin/Magnum Photos*
pp. 12–13 Jerusalem, Israel. *Patrick Zachmann/Magnum Photos*
pp. 14–15 Milan, Italy. *BlueRed/Keystone*
pp. 16–17 Cap d'Agde, France. *Yann Arthus-Bertrand/Corbis*
pp. 18–19 Srinagar, Jammu and Kashmir, India. *Steve McCurry/Magnum Photos*
pp. 20–21 Porbandar, Gujarat, India. *Steve McCurry/Magnum Photo*
pp. 22–23 Germany. *Adam Woolfitt/Corbis*

pp. 32–33 Polar bear in the Arctic. *Tim Davis/Corbis*
p. 34 Reindeer herd, Chukotka, northeastern Russia. *ArcticPhoto/Keystone*
A polar bear mother leads her cubs on Arctic sea ice. *ArcticPhoto/Keystone*
p. 35 Northern Fulmars off an iceberg near the coast of Antarctica.
ArcticPhoto/Keystone
Gentoo penguins, Antarctica. *DLILLC/Corbis*
pp. 36–37 Tombstone Valley, Yukon, the Canadian Arctic.
Bryan and Cherry Alexander/Keystone
p. 38 Brown bear, Katmai National Park, Alaska, USA. *Roy Corral/Corbis*
Meadow with purple loosestrife and meadowsweet, Scotland, United Kingdom.
Stuart Franklin/Magnum Photos
p. 39 Dworshak State Park, Idaho, USA. *ZUMA/Dukas*
Barred owl, Virginia, USA. *Wally McNamee/Corbis*
pp. 40–41 Wildebeest, savannah in Kenya. *Arthur Morris/Corbis*
p. 42 Maquis shrubland, Spain. *Markus Schweiss/Wikimedia Commons*
A Bengal tiger in Bandhavgarh National Park, Madhya Pradesh, India.
Mark Newman/Keystone
p. 43 Sand dunes and a rain-fed freshwater lake near Oursi, Burkina Faso.
Georg Gerster/Keystone
The Thar Desert near Jaisalmer, Rajasthan, India.
Keren Su/Lonely Planet Images/Keystone
p. 44 Rainforest in the Danum Valley, Borneo, Malaysia. *Frans Lanting/Corbis*
A jaguar, an iconic animal of the Central American rainforest.
Michael Fogden/Keystone
p. 45 Helmeted iguana in a rainforest in Costa Rica. *Michael Fogden/Keystone*
Congo Rouge mountain rainforest on Mahé Island, Seychelles. *Röder/Keystone*

CLIMATE HISTORY

BETWEE
FIREBAL
ICE DESE

Christian Rentsch

N

L AND

ERT

We cannot understand climate change without first knowing about the past history and evolution of climate. This is because climate changes can be verified only by comparing past conditions with present conditions. How was it then, how is it now? How will it be in a hundred years?

The fact that Earth's average annual temperature is currently about 15 degrees Celsius (59°F) may be a rather unspectacular piece of information. Knowing it is 0.7 degrees (1.3°F) higher than it was one hundred years ago is more impressive, and comprehending that average temperatures in the Northern Hemisphere are warmer today than at any other time in the past 1,300 years can make us sit up and start paying attention.

The paces and rhythms of climate processes vary widely. Some changes progress over a few years or decades while others take hundreds, thousands, or millions of years; some recur more or less regularly while others take us by surprise, occurring at unpredictable intervals or happening only once. They tell us a never-ending and truly unique story.

The study of Earth's climatic history, called paleoclimatology, is real detective work. Reliable weather data have been recorded for little more than 150 years. The oldest written records date back several thousands of years. The farther we go back in the history of Earth, the fainter the traces become, although sediments, ice cores, growth rings in petrified trees, and similar findings do provide us with a wealth of information. Increasingly sophisticated analytical instruments and chemical methods such as isotope measurement today provide climate scientists with evidence allowing them to discover with ever greater precision what kind of climate conditions prevailed on Earth in past ages. Thanks to these findings, called climate proxies or proxy data, which can be improved by adding information

from physical calculations of parameters such as solar activity or the Earth's orbit, we now have a rudimentary yet more or less conclusive understanding of the history of Earth's climate. Nevertheless, some conclusions are still based on guesswork and speculation.

Paleoclimatology is one of many keys to understanding current climate changes and it serves as a point of reference for predicting the climate of the future. The more precise the scientific understanding of climate history is, the easier it becomes to appreciate the many factors contributing to climate change and their complex coupling and feedback mechanisms. The more accurately the natural causes of past climate changes can be deduced from the study of climate history, the more accurately we can assess how anthropogenic influences—influences generated by human activity—could affect the Earth's environment and climate.

Yet there can be no certainty. This is not due to gaps in scientific knowledge, but rather because the human-climate system is by nature a deterministic, chaotic system. Even if each individual process were completely comprehensible and governed by predictable deterministic laws, it would still be impossible to forecast how the system would behave in its entirety.

Continents are drifting. At the San Andreas Fault in California, the Pacific Plate is moving northward relative to the North American Plate at a rate of one centimeter per year. *Shattil and Rozinski/Keystone*

Verrucano, a succession of red claystones, sandstones, and conglomerates, formed 270 million years ago when the Variscan mountains eroded in an extreme monsoon climate. Glarus Alps, Switzerland. *Helmut Weissert/ETH Zurich*

Black marls, rich in marine organic matter, alternating here with limestone, were formed in a greenhouse climate about 100 million years ago in an oxygen-deficient ocean. Southern France. *Helmut Weissert/ETH Zurich*

Limestone formations in the Swiss Alpstein massif are witnesses of a tropical coastal sea that existed here 130 million years ago under pronounced greenhouse climate conditions. *Alessandro Della Bella/Keystone*

Dolomite rock formed 220 million years ago in a shallow sea that extended over large parts of Southern Europe. Climate conditions were similar to those of the present-day Persian Gulf. *Helmut Weissert/ETH Zurich*

Many landscapes have been shaped by glaciers. Tyrolerfjord in northeastern Greenland illustrates how ice carves a huge valley from solid rock. *Simon Fraser/Science Photo Library/Keystone*

Fossilized sea lilies in marine sediment roughly 150 million years old; these animals grew at the sea bottom and displayed long, feather-like arms. Dorset, England. *George Bernard/Keystone*

Petrified fern preserved in slate. Minnesota, USA. *Layne Kennedy/Corbis*

Fossilized Neuropteris and Alethopteris, gymnosperms (naked seed ferns) from the Permian period nearly 300 million years ago. Found in North America. *Theodore Clutter/Keystone*

Fossil of the crow-sized Archaeopteryx lithographica, one of the oldest known birds. A warm climate some 150 million years ago fostered the evolution of large animals, including saurians. *James L. Amos/Keystone*

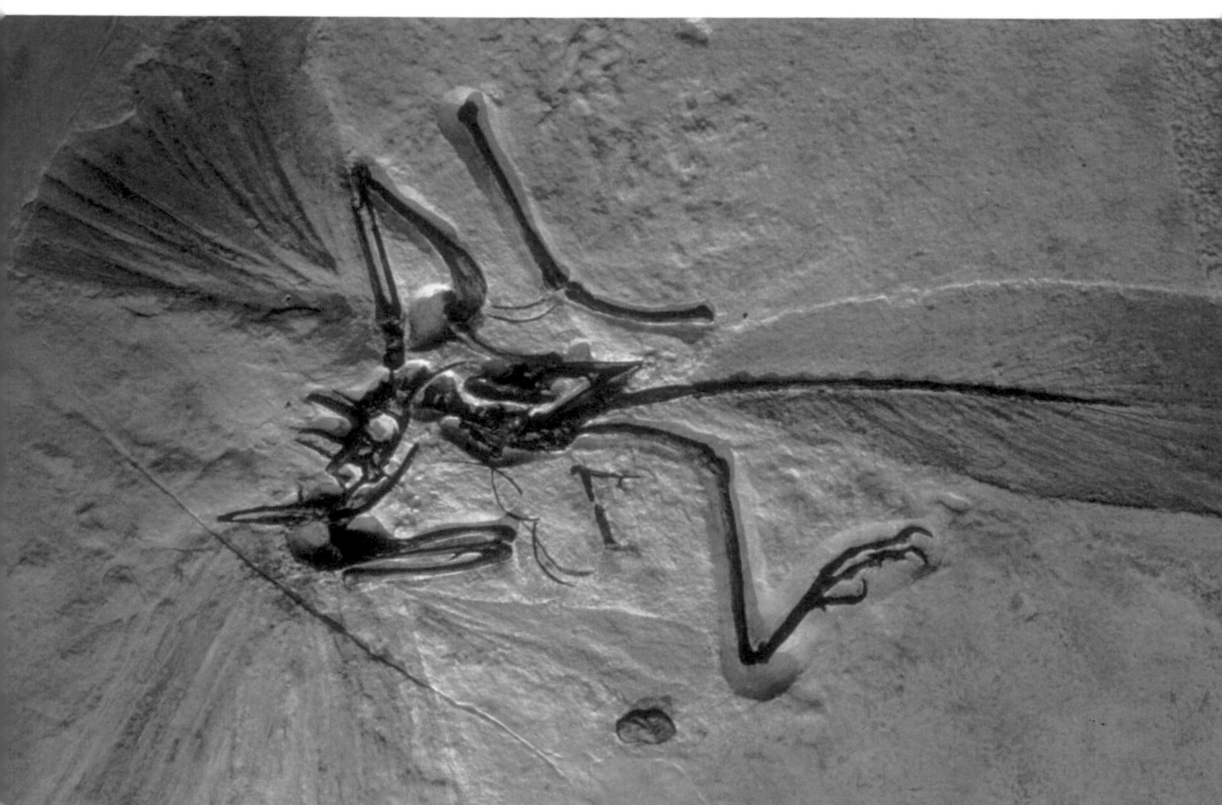

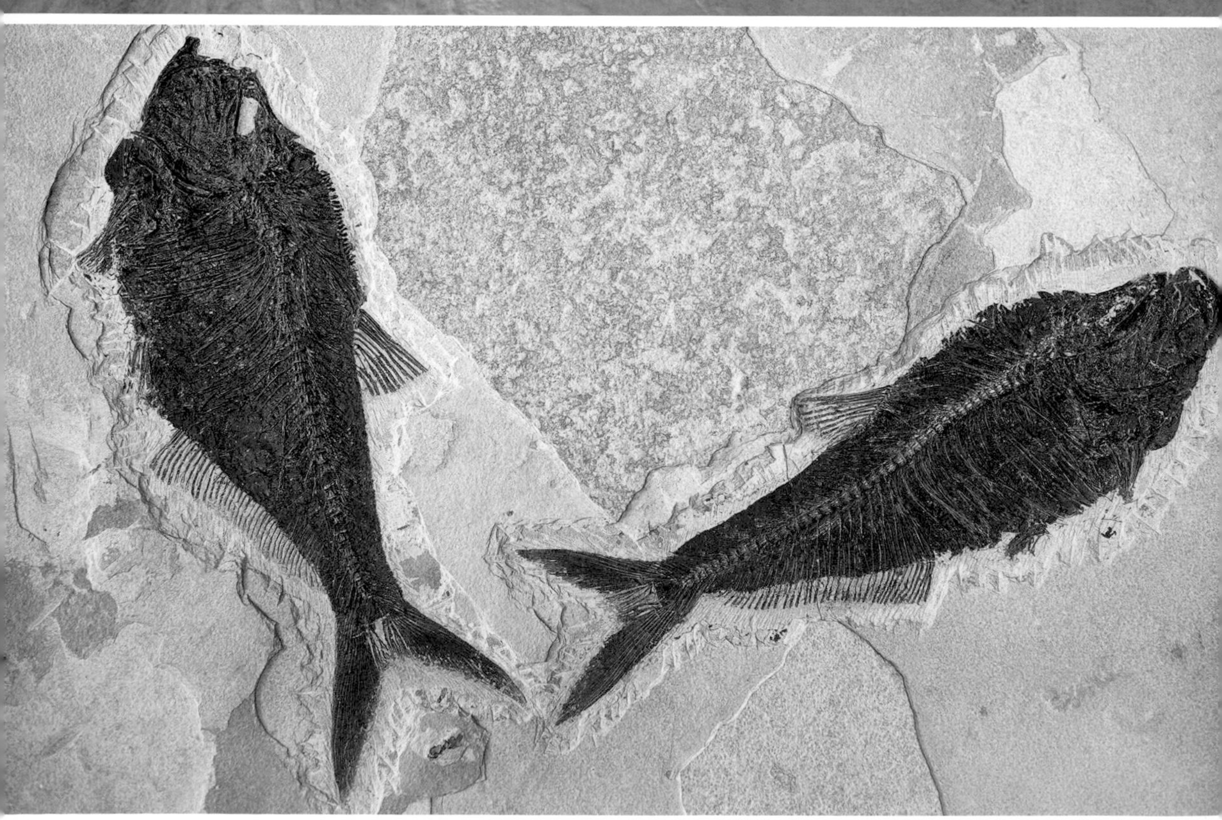

Fossilized Diplomystus from Wyoming, USA. These fish lived in the warm and humid climate of the Eocene epoch, were about forty-five centimeters long, and resembled modern-day herrings. *Theodore Clutter/Keystone*

A 460-million-year-old trilobite from Québec, Canada; this marine animal had an external skeleton and populated the seas starting in the Cambrian revolution 542 million years ago. *Marc A. Schneider/Keystone*

Crayfish fossils are rare because of their decomposable shells. Finds from the Cambrian period more than 500 million years ago prove that they were among the first animals to live on Earth. *Frans Lanting/Corbis*

Saber-toothed tigers, cave bears, and mastodons became extinct within a short time some 10,000 years ago. The causes could have been sudden climate change or human interference. *Bettmann/Corbis*

HOW THE EARTH GOT ITS CLIMATE

Climate history has been the history of the Earth. Whether continents have collided and broken apart, or new oceans and mountain ranges have emerged, these events have always had a profound effect on climate. In turn, changes in climate have also changed the face of Earth. Steppes have given way to huge tropical forests, warm and humid climate zones have turned into ice or sand deserts, entire areas have disappeared into the sea, and oceans have become inland seas or dried out completely. In short, Earth's history and climate history are interconnected in a multitude of extremely complex ways and have great influence on each other.

Although daily weather events occur mostly in the lower part of the atmosphere, climate changes, meaning changes in average weather patterns over a prolonged period of time, are processes involving the entire atmosphere as well as other realms of Earth—its mantel, oceans, ice and snow masses, and the entire biosphere. And of course the sun too, in providing our planet with energy, affects terrestrial climate patterns.

Earth did not always have an atmosphere. When cosmic dust clumped together about 4.6 billion years ago to form our planet, it also attracted light and inert elements, gases such as hydrogen and helium, methane and nitrogen, ammonia and hydrogen sulfide. But solar winds, strong plasma currents of electrons and protons, soon blew these gases away from Earth back into space.

It took about 600 million years for the Earth's mass to increase and its gravitational pull to become strong enough to hold these inert gases. This first atmosphere was very different from today's air. It was a toxic cloud of water vapor (80 percent), carbon dioxide (10 percent), hydrogen sulfide (5 to 7 percent) and other gases such as nitrogen, carbon monoxide, methane, and ammonia. In comparison, the two dominant gases at the time, water vapor and carbon dioxide, now together make up less than 1 percent of our air. There was no oxygen in Earth's first atmosphere—today the oxygen content of air is about 21 percent.

In many other respects, this primeval planet hardly resembled Earth as we know it today. It took more than half a billion years for the planet, a hot fireball of molten rock, to cool down enough for a solid crust to form around its core, which was several thousand degrees in temperature. Another several hundred million years passed before the temperature dropped well below 100 degrees Celsius (212°F) and water vapor condensed and turned into rain, reducing the amount of water vapor in the air. Today the global average content of atmospheric water vapor is barely a fourth of 1 percent.

Once this happened, two important driving forces of climate change were set into motion: the water cycle and the carbon cycle. They control the Earth's climate, regulating the greenhouse effect, and distributing solar thermal energy around the globe and through its oceans. These cycles are wheelworks linking all processes on Earth, from the biochemical processes found in every living creature, no matter how small, to worldwide processes such as ocean circulation, making all of them part of one large global mechanism.

Collectively, over a period of two billion years, these two cycles have been responsible for successively lowering the concentration of carbon dioxide (CO_2) in the atmosphere down from 10 percent. For the past 350 million years at least, CO_2 concentrations have stayed in a narrow band of 0.2 to 2 parts per thousand (per mil) or—in climatological language—between 200 and 2,000 parts per million (ppm).

This decrease was generated by a geochemical reaction that always occurs when rock weathers, a process which draws carbon dioxide molecules from the air. These molecules react with rock minerals to form bicarbonate, a water-soluble form of carbon dioxide. Streams and rivers carry the bicarbonate to lakes and oceans where it is deposited and permanently stored in sediment as carbonate.

The weathering of rock has been taking place for billions of years, and occurs wherever rock comes into contact with water. Indeed, all carbon dioxide would have been withdrawn from the atmosphere long ago if phenomena like volcanism had not continuously counteracted this process. Movement in the Earth's core forces some sediment to shift into the deeper hot and molten layers of the Earth's mantle, where it melts. This melting process transforms the bound carbonate back into carbon dioxide, which is returned to the atmosphere when magma erupts from volcanoes.

But where does oxygen, nonexistent in our primeval atmosphere, come from? Oxygen is produced almost exclusively by living organisms. To this day we still don't know exactly when and how life on Earth originated. The oldest traces of microorganisms, cyanobacteria (also known as blue-green algae), have been found in rock layers as old as 3.8 billion years. Like nearly all later plant organisms, they obtain their energy through photosynthesis, a biochemical process that uses sunlight to convert carbon dioxide and water into carbohydrates, releasing oxygen in the process. This shows that very early in the history of Earth living organisms began influencing the climate.

But it took some three billion years for the concentration of oxygen in the air to increase markedly. In the beginning, all the oxygen produced by blue-green algae was immediately bound to iron molecules in water. Only after almost all the iron in the oceans had oxidized did oxygen gas begin to be released into the atmosphere. Even as recently as 1.4 billion years ago, the concentration of oxygen in the air is assumed to have reached a mere 0.2 percent. Recent studies indicate that oxygen levels at the beginning of the Paleozoic era some 542 million years ago fluctuated between 15 to 20 percent; about 300 million years ago the concentration of oxygen even exceeded 30 percent, a much higher level than today's. Within a few million years at the beginning of the Mesozoic era, oxygen levels then dropped again by almost a third. Climate scientists believe this was caused by the massive deforestation of coastal and fluvial regions due to flooding. It was not until about 25 million years ago that the concentration of oxygen in the air stabilized and reached an average level of about 21 percent, and this has stayed more or less the same since then.

ICE AGES AND WARM PERIODS: GREENHOUSE GASES AND DRIFTING CONTINENTS

The sun has been Earth's only significant source of heat since the original fireball cooled down and became a planet. Earth's core, which is still extremely hot, does not affect temperatures on the surface of the planet.

In 1879, physicists Josef Stefan and Ludwig Boltzmann drew attention to a curious fact. According to their calculations, direct sunlight alone would suffice to warm up the planet's surface to only minus 18 degrees Celsius (0°F), a very cold and hostile temperature. But Earth's average temperature is in fact about 15 degrees above zero (59°F). How can we explain this difference of 33 degrees (59°F)?

The answer to this mystery is found in the natural greenhouse effect resulting from a special property inherent in some atmospheric gases, now generally referred to as greenhouse gases. They have the capacity to absorb large amounts of Earth's thermal radiation, thus retaining heat in the atmosphere. They work somewhat like the glass panes of a greenhouse, which let sunlight through but prevent the sun's heat from escaping the greenhouse.

Greenhouse gases include water vapor (H_2O), carbon dioxide (CO_2), methane (CH_4), ozone (O_3), and other trace gases. Although these gases together do not make up more than half a percent of our atmosphere, their greenhouse effect generates an increase in temperature of about 33 degrees Celsius (59°F).

The most important greenhouse gas is water vapor—water in its gaseous state. It contributes roughly two-thirds to the natural greenhouse effect. In spite of the important role it plays, it is not a matter of greater concern in the climate debate because water has a special physical characteristic: it evaporates exactly within the Earth's temperature range. Thus air humidity, meaning the concentration of water vapor in the air, is dependent on air temperature, making water vapor very unlike other greenhouse gases. In other words, any rise in temperature increases the content of water vapor in the air, which in turn drives up temperatures even further.

The concentration of the greenhouse gases carbon dioxide and methane is governed by biogeochemical processes, such as rock weathering or the growth of algae. Unlike air humidity, which fluctuates depending on time and place, these greenhouse gases are more or less evenly distributed across the planet and remain in the atmosphere for many decades. Although their levels of concentration in the air are much lower than those of water vapor, CO_2 and CH_4 combined are responsible for about one-third of the greenhouse effect.

Climate scientists have been able to deduce from indicators called proxy data →p.100 that greenhouse gas concentrations fluctuated considerably throughout Earth's early history. The specific reasons for these fluctuations are a matter of debate among scientists. However, we do know for a fact that the almost total depletion of carbon dioxide due to rock weathering and the growth of cyanobacteria, or blue-green algae, repeatedly reduced the greenhouse effect so drastically that most of the Earth froze over for periods of 200 to 300 million years. The planet was covered by a layer of ice several hundred meters thick, but scientists contest whether the surface was entirely frozen, as suggested by the Snowball Earth hypothesis.

These ice ages ultimately ended because of a control mechanism working like a thermostat. Underneath thick layers of ice, the depletion of CO_2 through rock weathering and blue-green algae came to an almost complete standstill, while volcanoes continued to emit large amounts of carbon dioxide and methane. In this way, the greenhouse effect started up again over the course of millions of years. The ice masses melted and the Earth's climate grew warmer—possibly, as some proxy data suggest, reaching temperatures as high as 50 degrees Celsius (122°F).

These warm temperatures triggered an extraordinarily intense growth spurt 542 million years ago at the beginning of the Cambrian period. One of the causes of this rise of life, known as the Cambrian Explosion, was what we refer to as continental drift.

In a groundbreaking study published in 1915, geophysicist Alfred Wegener first proposed that the Earth's crust is not one static shell of rock. But nearly fifty years went by before science found an explanation for the drifting of continents. Current plate tectonics theory assumes that the Earth's crust consists of several plates, each 70 to 100 kilometers thick. These plates float on the Earth's soft, viscoelastic upper mantle and are in constant motion, breaking apart and drifting away from each other, or colliding with and sliding one over the other. Their rims are characterized by violent volcanic eruptions and earthquakes, and even underneath the oceans, magma from deep inside the Earth rises up, pushing plates apart or over one another.

Shortly before the beginning of the Cambrian period 542 million years ago, a huge supercontinent broke apart into several smaller continents. New oceans and ocean currents emerged between these continents, setting vast masses of water into motion. New, mineral-rich coastal waters provided marine animals and plants with good living conditions and encouraged reproduction.

But it is very likely that this stark increase in marine life was precisely what caused the first great mass extinction in Earth's history 50 million years later, as marine organisms unwittingly destroyed their own habitat, the basis of their very existence. Their constantly growing populations increased the need for carbon, which lowered concentrations of CO_2 in the atmosphere. This in turn weakened the greenhouse effect, and large areas of the Southern Hemisphere started to freeze over again. Sea levels dropped by up to 100 meters, extensive areas of coastal waters dried up, and numerous plant and animal species died out. This mass extinction at the end of the Cambrian period was responsible for the disappearance of more than half of the planet's existing marine species.

Meanwhile however, marine plants had released so much oxygen during the Cambrian Explosion that a thin ozone layer formed in the stratosphere, surrounding the Earth's surface at a distance of about 15 to 50 kilometers. (The ozone layer today still absorbs about 70 percent of the sun's energy-rich UV radiation that is deadly for any form of life.) For the first time, life was able to evolve outside the oceans too—a milestone in climate and natural history. As plants spread across land surfaces, vegetation became a new climate factor.

In the 160 million years following the Cambrian Explosion, terrestrial plants conquered every continent. Humid tropical zones were covered with huge forests of giant ferns up to forty meters high, tree-like horsetail plants and club mosses. (Their remains today make up Earth's extensive anthracite coal deposits.) Fifty million years later, approximately 300 million years before our time, the first animals began to populate land in coastal regions.

By the end of the Mesozoic era some 65 million years ago, an enormous number of different animal genera had appeared, forerunners of almost all of today's animal species. The waters were inhabited by sponges, corals, mussels, fish, crustaceans, turtles, snails, and amphibians; the land was populated by insects and the first vertebrates—snakes, lizards and crocodiles, birds, and dinosaurs.

Yet this long evolution of animals and plants did not always go smoothly. During the Earth's history from 360 to 65 million years ago, there were four mass extinctions. A major share of life fell victim each time, with up to 85 percent of living creatures dying off.

Theories on the causes of these mass extinctions are widely diverging. But there is no question that climate changes brought about by continental drift and related volcanic eruptions played an important role in these events →p. 70.

In the Paleozoic era, for instance, four smaller continents united in the Northern Hemisphere to form one large northern continent now known as Laurasia. It drifted toward the South Pole where it collided with Gondwana, the southern continent. As a result, Pangaea, a single supercontinent, took shape at the end of the Paleozoic. This continent extended from the south pole over all latitudes to the far north. Crushing pressure in collision zones caused the plates to buckle upward, creating great mountain systems. The Appalachians, the Vosges, the Black Forest, and the Urals are remnants of those mountains.

At the beginning of the Mesozoic some 250 million years ago, Pangaea divided into a northern and a southern continent. To the east, the Tethys Ocean squeezed itself between the two continents. Today, the Mediterranean Sea is what is left of that ocean. In the Northern Hemisphere, the North American plate gradually broke away from Eurasia and the future North Atlantic began to form. In the Southern Hemisphere, South America and Africa drifted apart.

These massive shifts were accompanied by drastic climate changes, and some of them have yet to find full explanation. The weakening of the greenhouse effect and the subsequent cooling of the climate could have played a role; increased volcanic activity, especially in times of intense tectonic movement, could have led to an extreme greenhouse climate. Not least, huge amounts of the poisonous gases released by volcanic eruptions might have wiped out a large share of Earth's living organisms or damaged the ozone layer that protected them from UV radiation.

But between these cataclysmic events, the climate was temperate, drier, and several degrees warmer than it is today, even in higher latitudes. Extensive tropical and subtropical forests of conifers, giant sequoias, ginkgos, and palm trees characterized river valleys and lakelands, especially in Europe and North America. Animal life was dominated by many different types of lizards, notably dinosaurs and pterosaurs (Greek for winged lizards). Their more developed anatomy made them far superior to the older crawlers. Thanks to their agility, they were able to conquer the entire supercontinent of Pangaea before it broke apart, which explains why traces of dinosaurs can be found in Africa, Europe, Asia, and North and South America.

We still do not know why dinosaurs became extinct at the end of the Mesozoic era 65 million years ago. In 1980, a team of researchers working with Nobel laureate physicist Luis W. Alvarez proposed a hypothesis based on the analyses of sedimentary layers. They suggested that the powerful impact of a meteor could have wiped out dinosaurs. A gigantic impact crater 180 kilometers in diameter and 900 meters deep, dating from precisely that period, was indeed discovered buried under thick layers of sediment in 1991 in the Yucatán peninsula in Mexico. Calculations showed that this impact released five times the amount of energy in all of today's nuclear weapons combined. The power of the blast from the impact would have wiped out all life in a radius of about 1,000 kilometers. Red-hot rocks ejected by the blast would have ignited huge forest fires within an even greater radius. Enormous amounts of dust and other particles could have blocked sunlight and darkened the entire planet for months or even years, leading to a sudden and very drastic drop in temperature.

The reasons behind an extremely warm phase that occurred 55 million years ago during the Cenozoic era, when temperatures rose to levels never seen again, have not been fully explained either. Average temperatures in the tropics rose to more than 38 degrees Celsius (100°F), and even in the Arctic, temperatures were as high as 20 to 25 degrees (68–77°F). This thermal maximum at the boundary between the Paleocene and Eocene epochs, referred to by climate scientists as PETM (Paleocene-Eocene Thermal Maximum), was possibly caused by powerful movements and landslides in ocean basins, which could have triggered the release of large amounts of carbon dioxide and methane gas. Methane is stored in the cold, deep regions of the oceans as frozen methane hydrate, and hot springs, magma, or landslides can cause its violent and abrupt release.

Average temperatures on Earth have never been as high since then. The gradual lowering of the temperature by about 7 degrees Celsius (12.6°F), which took place over the next 20 million years, was mostly due to a complex process of interaction between the oceans, the atmosphere, and continental land masses.

CONTINENTAL DRIFT BETWEEN THE PERMIAN AND THE QUATERNARY

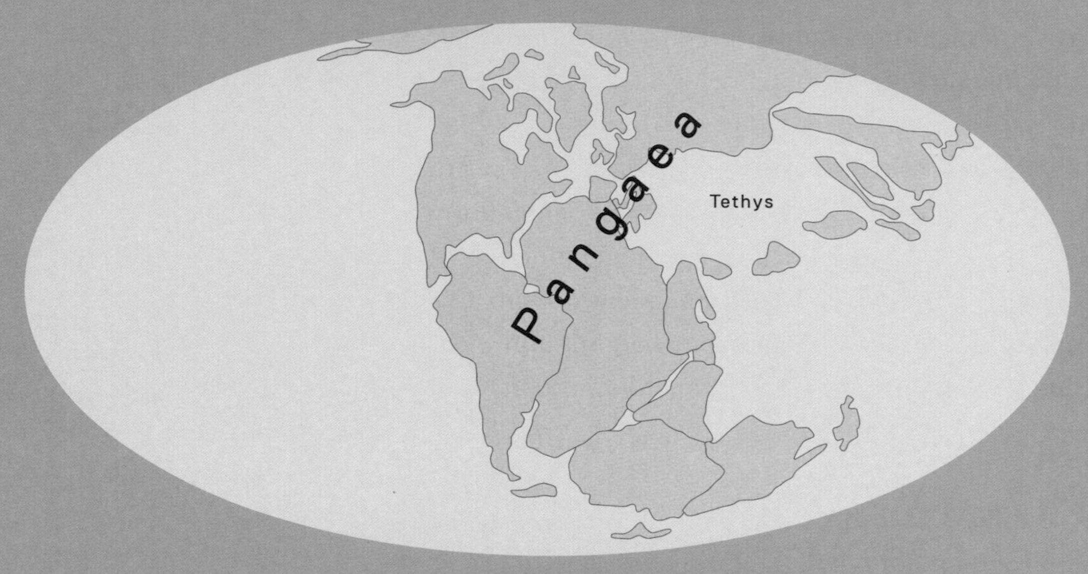

Permian
from 299 to 251 million years ago

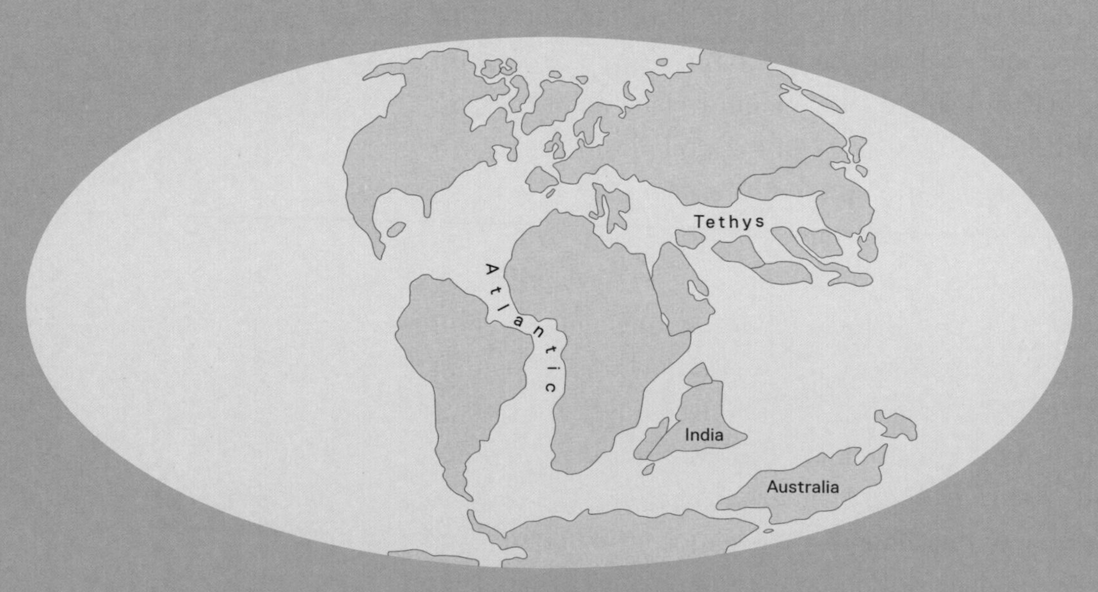

Jurassic
from 199 to 145 million years ago

Sources: Hermann Schäfer, Forschungsinstitut und Naturmuseum Senckenberg, Frankfurt am Main; Northern Arizona University, 2008

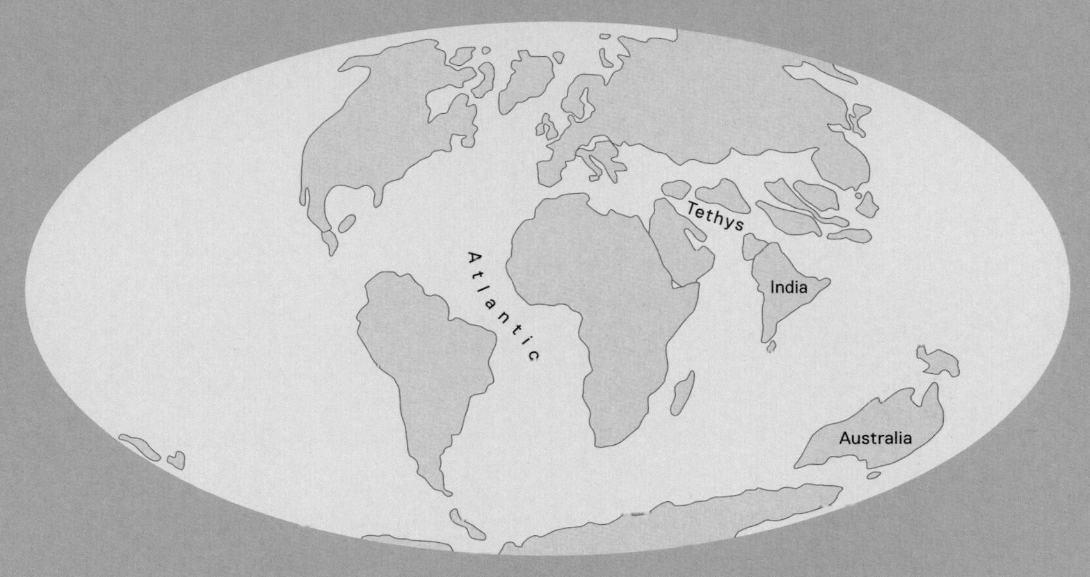

Tertiary
from 65 to 1.8 million years ago

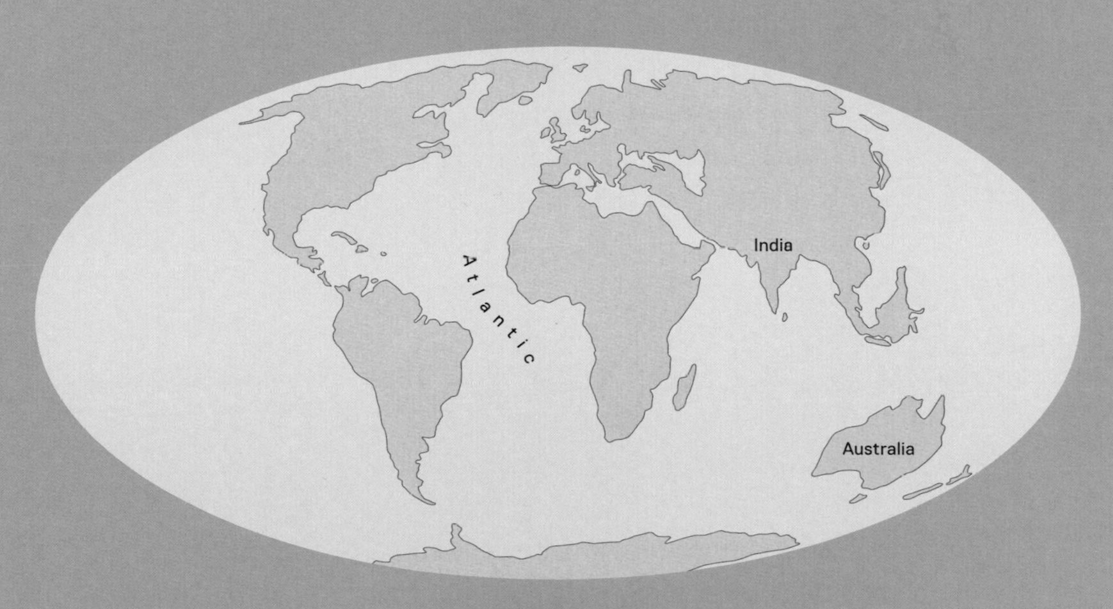

Quaternary
from 1.8 million years ago to present

HOW CONTINENTAL DRIFT CAUSES CLIMATE CHANGE

Some 34 million years ago, the Earth cooled down noticeably. The climate has never since come close to experiencing the average temperatures that had previously prevailed for more than 100 million years. In other words, the planet's climate stabilized at a new temperature level approximately 6 to 7 degrees Celsius (10.8–12.6°F) lower.

One of the reasons for this lasting cooling down was a tectonic development in Antarctica, starting about 70 million years ago, when present-day Australia separated from the ancient southern continent of Gondwana. Some 35 million years later, the last remaining land connection to the west between Gondwana and the westward drifting South America disappeared. Antarctica became an island completely surrounded by the Southern Ocean.

These continental drifts had far-reaching consequences for the further development of the planet's climate. The Earth's rotation set into motion a powerful cold-water gyre around Antarctica driven by strong polar winds. Known as the Antarctic Circumpolar Current, this gyre has since then isolated Antarctica from warmer surface currents coming from the tropics. It also connected the ocean currents of the Atlantic, the Pacific, and the Indian Ocean, forming a single, global ocean current called the global ocean conveyor belt.

The mixing of these ocean currents with ice-cold polar water constantly cools deep waters before they turn back from the Southern Ocean toward the Equator. In geological eras preceding the development of this circulation system, temperatures in deep ocean currents easily exceeded 8 degrees Celsius (46.4°F). Since the system's formation, temperatures have never again risen much over 4 degrees Celsius (39.2°F). The Antarctic Circumpolar Current thus works like a huge, global cooling unit.

Moreover, this cold water barrier turned Antarctica into the Earth's largest cold-air reservoir. Even though temperatures in summer rose several degrees above the freezing point for millions of years and extensive forests once covered the continent's coastal areas, Antarctica has never again been totally free of ice. Quite the contrary—it has been completely covered by a layer of ice up to 4.5 kilometers thick for at least the past five million years.

Another factor contributing to the Earth's cooling was a sustained decrease in greenhouse gases. While CO_2 levels during the entire Mesozoic era were generally six to nine times higher than today, they have not substantially exceeded present-day levels for at least 20 million years.

To explain this drop, scientists suspect that diatoms, microscopic marine algae with a hard silica cell wall (skeleton), could have played a major role in keeping CO_2 levels down. Like blue-green algae, diatoms absorb carbon dioxide and bicarbonate from water. When blue-green algae die, most of the carbon dissolves back into the water. This is not the case with diatoms. When they die, their skeletons sink down to the ocean floor, taking considerable quantities of carbon with them. Over the course of millions of years, large amounts of carbon have in this way been deposited in layers of sediment made up of diatom skeletons.

Sediment analyses prove the global ocean conveyor belt has in fact greatly accelerated diatom reproduction. Today, diatoms absorb 100,000 to 200,000 tons of carbon from the carbon cycle every year.

Continental drift continued to influence the way climate evolved, as for instance when the Fram Strait opened between Greenland and Spitsbergen about 23 million years ago. The Arctic Ocean was originally a landlocked body of water encompassed by Eurasia and North America, and indeed may even have been a freshwater sea. The opening of the Fram Strait connected it to the global ocean conveyor belt. As a result, warmer water from the tropics carried more humidity to the northern polar region, generating an increase in precipitation, especially more snow in winter. In turn, cold water from the Arctic Ocean contributed to a steady cooling of the North Atlantic deep ocean current. These conditions still prevail today.

The tectonic folding that created a land bridge between North and South America also greatly altered ocean circulation in the North Atlantic. The flow of water from the Pacific to the Atlantic was interrupted. The impact of this event was strongest on the opposite side of the Atlantic some 7,000 kilometers away. Water from the Atlantic no longer mixed with water from the Pacific, which is less salty. Since a higher salt content makes water heavier, warm and saltier water from the Caribbean, transported north through the Atlantic by the Gulf

Stream, sank to a much deeper level when it reached the coast of Greenland. This generated a suction effect that strengthened the Gulf Stream and in turn fostered Europe's mild climate. At the same time, the warm Gulf Stream brought more humidity to cold north polar regions. The consequent increase in precipitation, mainly as winter snowfall, caused the climate in the polar region to become cooler instead of warmer, in stark contrast to the situation on the European continent.

Continental drift has influenced our climate for more than 320 million years, and it will continue to do so because continents have not stopped moving. But this drift has hardly any appreciable influence on climate development within a time frame of a few million years.

CLIMATE ON A ROLLER COASTER RIDE

The Earth has been in an ice age for the past 2.6 million years. This sounds like a paradox, because climate scientists say we have been living in a warm period for the last 10,000 years. But this is just a matter of definition. In climatology, the term ice age—not to be confused with glacial period—refers to periods in the history of Earth in which both poles are covered by ice sheets. This happens much less frequently than one would normally assume. The poles were largely free of ice throughout 80 to 90 percent of Earth's history. Within an ice age, climatologists differentiate between cold periods or glacials and warm periods or interglacials.

Whatever may have caused the vastly different climate conditions belonging to various geological eras, periodic fluctuations within individual eras must have had other causes.

More recent climate models support the assumption that fluctuations between glacials and interglacials during the last few hundred thousand years have been generated by more than a single factor. It seems that the somewhat irregular ebb and flow of cold and warm periods is triggered by several periodic variations of varying strength and duration; these variations can overlap, reinforce, or weaken one another.

Important among these periodically occurring cyclic variations are the Milanković cycles, named after the astronomer Milutin

Milanković who discovered them →p.124. His theory makes a connection between the episodic nature of warm and cold periods and variations in the parameters of the Earth's orbit. These variables comprise changes in the Earth's orbit around the sun, the Earth's axial tilt, and the wobbling movement of the Earth's axis. Together they are responsible for slight variations in the total amount of solar energy reaching the Earth, but more importantly, they also play a role in the distribution of solar energy across different latitudes of the Northern and Southern Hemispheres. Because these three astronomic cycles of varying frequency overlap, the Milanković cycles of the past three million years paint a somewhat turbulent picture, with marked highs and lows at more or less regular intervals that closely coincide with glacial and interglacial periods. During the first 1.8 million years, the long waves in the Milanković cycles spanned about 41,000 years; in the last 850,000 years they have spanned about 100,000 years. The cold periods lasting some 90,000 years each were substantially longer than the warm periods that lasted only 10,000 years.

Until recently, the Milanković cycles were considered to be the most plausible explanation for fluctuations in climate over millions of years. But most climatologists today believe that the impact of these astronomical factors on the transition between cold and warm periods has probably been overestimated. If the climate system itself—in the absence of external drivers—is able to generate short-term changes such as the periodically recurring El Niño phenomenon, why shouldn't other fluctuations exist that occur over hundreds of years as well?

The Atlantic multidecadal oscillation (AMO) supports this idea of "internal variability." AMO refers to fluctuations in water temperatures of the North Atlantic over a period of seventy to eighty years that cannot be explained by external factors or by an increase in the greenhouse effect due to anthropogenic (man-made) causes. AMO data do not reach back far enough to provide us with conclusive results, and what causes these variations is still far from clear. Computer simulations however show an astonishing conformity of AMO data with climate changes in the Northern Hemisphere over the last 150 years. We also know of numerous other feedback mechanisms, still needing more research, that are generated by the climate system and that show an effect only after many decades or even centuries.

In addition to these more or less cyclical fluctuations in the current ice age, there have always been climate events that do not fit in with periodic transitions between warm and cold periods. Since the 1980s, thanks to findings from ice cores, we have been able to accurately date climate changes to within a few decades. They show us that in the last 110,000 years alone, there have been twenty-three sudden global temperature increases. In each of these cases, temperatures in Greenland rose by up to 15 degrees Celsius (27°F) within just ten to twenty years and remained at that level for several centuries.

These sudden climate swings are called Dansgard-Oeschger events (DO events), named after their discoverers, paleoclimatologist Willi Dansgaard and climatologist Hans Oeschger. These events have yet to be fully explained, although they are most likely the result of abrupt changes in the ocean currents of the North Atlantic. If and how DO events are related to Heinrich events, which are abrupt drops in temperature, is also unclear. Named after oceanographer Hartmut Heinrich, they lasted around 750 years and took place at irregular intervals of several thousands of years. They were probably triggered by huge masses of ice breaking off the North American glacier. The weight of a continental glacier more than three kilometers thick would have caused its bottom layers to became unstable and set it in motion. Huge blocks of ice shelf broke off and drifted down the present-day Hudson and St. Lawrence Rivers into the North Atlantic. The melting of this ice released vast amounts of cold freshwater into the sea, which is believed to have disrupted Atlantic ocean currents for prolonged periods.

Two of the coldest glacial periods of the current Quaternary Ice Age were the Mindel or Elsterian glacial stage, which began 400,000 years ago and lasted around 80,000 years, and the Würm or Vistulian glacial stage, which began 117,000 years ago and lasted over 100,000 years. Continental glaciers up to three kilometers thick covered vast areas of Asia, Europe, and North America during both glacial periods. In Germany, the solid ice sheet extended south as far as Hanover and in North America almost as far as New York and Seattle. In the Alps and the Himalayas, glaciers spread down into valleys. Sea levels around the world dropped by 50 to 100 meters, and during the Würm glacial stage sometimes by even as much as 130 meters.

The Bering Strait, the English Channel and many other straits dried up and became land bridges.

The Würm glacial stage was the coldest period in the Quaternary ice age; the average global temperature barely reached 11 degrees Celsius (52°F). Average annual temperatures in Germany were 3 degrees (37°F), and even in the warmest summer months average temperatures did not exceed 8 degrees (46°F). These cold periods were also felt in the Southern Hemisphere, although not to such an extent. Some 80 percent of the Southern Hemisphere is covered by ocean waters, much more than in the north, and oceans cool much more slowly than continents do.

The gradual transition to today's warm period began 18,000 years ago and proceeded at a very fast pace, unusual in terms of climate history. The global climate warmed up within only 8,000 years, with temperatures rising to the present average of 14 to 15 degrees Celsius (57–59°F).

However, during this warm period there was also a cold snap of unprecedented severity. During the Younger Dryas period 12,700 years ago, average temperatures in the Northern Hemisphere again dropped down to about 10 degrees Celsius (50°F) in just forty years. The timberline receded throughout the world. In Scandinavia, but also in present-day Canada and northern parts of the United States, tundra spread where vast coniferous forests had previously stood. Southwest Asia experienced a period of drought that lasted hundreds of years. This cold snap ended 1,100 years later, as abruptly as it had started. Within a few years, temperatures rose again by around 7 degrees Celsius (12.6°F), going 2 degrees (3.6°F) above the present mean of 15 degrees (59°F).

What caused this abrupt drop in temperature? Rising temperatures after the end of the Würm glacial period (termed Wisconsin glaciation in North America) melted large parts of North America's continental glacier. Meltwater accumulated behind its eastern outlet glaciers, forming an immense freshwater inland sea called Lake Agassiz, which was as large as the Caspian Sea is today. Once the glacial barriers began to also melt, it did not take long before Lake Agassiz drained into the North Atlantic through the Hudson Bay, the Great Lakes, and the St. Lawrence River. Huge amounts of freshwater were released, changing the course of the warm Gulf Stream by pushing it south. Europe's "heat pump" was turned off. The entire circulation system of the North Atlantic Ocean may have been disrupted in this period. Ocean circulation went back to normal only after Lake Agassiz was almost completely drained—and climate warming once again took its course.

DO events, Heinrich events, and the unexpected cold snap of the Younger Dryas period have increasingly turned the attention of researchers in recent decades to climate phenomena that can disrupt intricately interconnected cycles in surprising ways. This includes nonlinear developments and tipping elements that suddenly end, accelerate, interrupt, or even reverse climate changes within a few decades or centuries. These tipping elements also include the most unpredictable factor: the human race.

HUMANS AS A CLIMATE FACTOR

Compared to the age of Earth, the history of the human race is but an instant; it does not account for more than half a thousandth of Earth's history, even if we are generous with our data. In a thousand-page chronicle of the Earth in which time relations are faithfully described, the human race would have to be content with a couple of sentences on the last page.

It is difficult to determine exactly when humans became human. Although the oldest skulls belonging to human-like creatures, found in Chad and Kenya a few years ago, date back about six million years, the first humans who walked upright and used stone tools probably lived about two million years ago.

The ancestors of today's Homo sapiens left their homeland in eastern Africa for the first time 1.75 million years ago. In the hundreds of thousands of years that followed, they gradually populated Europe and Asia, settling in higher and higher latitudes. In the Würm glacial stage, or Wisconsin glaciation, when for several millennia a land bridge emerged between eastern Asia and North America where the Bering Strait is today, some adventurous tribes must have walked from Siberia to Alaska. They and their descendants proceeded to populate North America, continuing to South America 13,000 years ago by way of the Isthmus of Panama. It remains a mystery to this day why, of all our human-like ancestors, only Homo sapiens (who first appeared in Europe 40,000 years ago) survived in the end.

Earth has not changed much since we humans made our appearance. The location of continents and the composition of the atmosphere then were very much as they are now. Nearly every one of today's animal species, including mammals, had already spread across all continents except Antarctica. Greenhouse gases, varying only slightly at times, had also evened out at a level no longer posing any threat to life on Earth.

Still, the glacial periods of the Pleistocene epoch, with its predominantly cool and changeable climate, must have been an extraordinary challenge for plant and animal life. While most animal and plant species can survive only in certain climate zones or need a very long time to adapt to climatic changes, humans turned out to be highly adaptable to different climates, learning to survive in a great variety of environments. Humans owe this ability to the development of their practical intelligence and foresightedness; their capacity for using tools, building shelters, and making clothes; their skill in the handling of fire; and, finally, their ability to organize and work collectively. Thanks to these qualities, Homo sapiens today is the only form of life on Earth that is also in a position to endanger its own survival.

The warm Holocene epoch has lasted for the past 10,000 years, going through numerous climatic fluctuations. Regional effects were occasionally much more serious than changes in global temperature averages might lead us to believe. Global temperature fluctuations of 1 or 2 degrees Celsius (1.8–3.6°F) have at several times drastically altered living conditions for people in the Holocene. Studies show that in many cases changes in climate have also played a role in historical developments and upheavals, in human migrations and wars. They have led to technical and social accomplishments, wealth and profusion, but also famines and epidemics.

Naturally there is no way of knowing for sure to what extent the extremely cold climate of the Younger Dryas period and the transition to the present warm period prompted nomadic hunter-gatherers to settle down and become farmers and herders. But it was surely no coincidence that these communities evolved almost simultaneously in different parts of the world during this particular period of climatic transition. These regions included the Fertile Crescent, extending from present-day Iran and eastern Turkey through Iraq, Syria, and Jordan to Lebanon, Israel, Palestine, and Egypt; as well as the Indus River Basin of the Indian subcontinent; the Yangtze and Yellow Rivers in China; the Niger in West Africa; and areas along the South American coast.

In Europe, where the climate was cooler, this agrarian transition, called the Neolithic Revolution, did not take place until some 3,000 years later. It could have been that several species of animals important to hunters such as mammoths, cave bears, and woolly rhinoceros could not adapt to warmer temperatures and became extinct. At the same time the habitat of nomadic hunters and gatherers dwindled because melting inland glaciers caused sea levels to gradually rise by 120 meters and extensive coastal areas and lowlands disappeared under the waters. A warmer and wetter climate offered sedentary societies good conditions for planting crops and raising sheep, cattle, goats, and pigs. As the barren tundra receded to the north, it made way for open grasslands and forests of pine and deciduous trees.

The Holocene Climate Optimum, the warmest phase of the current Holocene period to date, also created favorable living conditions in other parts of the world. Rivers, lakes, and lush grasslands with perennial rivers existed in the Sahara, which was populated at the time, as were other regions of Africa, which have since become desert. Tropical rainforests spread in the Amazon, and in Pakistan and India the warm summer monsoon brought abundant rainfall.

The Holocene Climate Optimum, which lasted from around 6,000 to 3,000 years before our time, was interrupted by a cooler period that came on suddenly and lasted 1,600 years. It brought drier conditions to the Middle East, large parts of Africa, and eastern Asia. The Saharan savannah dried out, gradually turning into the desert we know today. Scattered pastoral tribes left their settlements and migrated to the banks of larger rivers—in Africa, the Nile and the Niger, and in eastern Asia, the Euphrates and the Tigris. They first practiced rain-fed agriculture on the banks and mouths of the rivers that periodically flooded. But soon, even this was not enough to feed growing populations. As a result, the first ancient civilizations and

“Because of these anthropogenic emissions of carbon dioxide, global climate may depart significantly from natural behavior for many millennia to come. It seems appropriate to assign the term ‘Anthropocene’ to the present, in many ways human-dominated, geological epoch.”

Paul Crutzen, Nobel laureate in chemistry, 2002

political systems arose in Mesopotamia and ancient Egypt, and on the banks of Asia's greatest rivers.

The Holocene Climate Optimum, so insignificant to climate history, is paramount to the history of mankind. The emergence of the first civilizations, large cities, and structured political systems laid the groundwork for the development of craft skills, arts, and sciences. Cities had to be supplied with food from near and far; irrigation systems were devised and agriculture was planned and organized collectively; economic systems, trade, and transportation developed. As a result, since then, population growth stopped being directly dependent on the fertility of local surroundings.

Even after the Holocene Climate Optimum had ended, the climate continued to fluctuate between warm epochs called *optima* and cold ones called *pessima*. These moderate climate swings also affected human living conditions significantly.

The Iron Age Cold Epoch from about 900 to 300 BCE was the coldest period in the Holocene, with temperatures 1 or 2 degrees Celsius (1.8–3.6°F) below current-day temperatures. It led to long-lasting periods of drought that affected the entire world. Numerous rivers dried up and in mountainous regions glaciers expanded far down into valleys. Food became scarce in many parts of Europe. Entire tribes from northern Europe migrated south. The most fertile regions of western India and Pakistan became steppes and the Thar Desert grew to the size of New Zealand.

The somewhat warmer temperatures of the Roman Climate Optimum, roughly from 200 BCE to 300 CE, very likely contributed heavily to the expansion of the Roman Empire. Alpine passes were open year-round, making it possible for trade to prosper between Rome and its northern provinces. France and Germany became the breadbaskets of the Roman Empire. Hannibal's legendary crossing of the Alps with his army of elephants would have not been feasible in a colder climate.

It was probably the arrival of the cold and arid steppe climate of the Migration Period Pessimum from 300 to 600 CE, and not the urge to conquer, that drove Asian steppe tribes to descend on China and India. To the west, they advanced through southern Russia and areas along the Danube all the way to France. The resulting exodus of Germanic tribes finally led to the end of the Roman Empire.

Average temperatures during the Medieval Warm Period or Medieval Climate Optimum, from 1000 to 1200 CE, rose close to present-day levels, at least in the Northern Hemisphere. An expansion of cultivated areas and higher crop yields is likely to have contributed to intensive settlement in the northern and eastern parts of Europe, and to the flourishing of medieval towns. Grains could be cultivated even on Iceland, enabling the Vikings to establish large settlements, which they did in Greenland as well.

The Little Ice Age, from 1250 to 1850, is more a period of very unsettled, cooler weather than a real ice age. For Europe it meant long, cold winters, massive storm floods in coastal areas, and rainy summers. The cold, epidemics, and famines claimed the lives of more than a million people. There were several waves of mass emigration from Central Europe to America. The population in Central Europe dropped by 30 to 40 percent.

What these climate swings have in common is that they were all evidently due to natural causes. Climatologists have plausible explanations for some of them: changes in the activity of the sun; increased volcanic activity; the East Greenland Stream, which for a few centuries was ice-free; and the internal variability of medium-term climate changes. But even today these fluctuations have yet to be explained in full detail. Some climate changes could have been generated by human activity. According to climatologist William Ruddiman, the forced cultivation of rice in Asia and animal herding could have altered methane concentrations in the atmosphere, although proxy data, climate records obtained from natural sources, do not seem to confirm this theory. Moreover, ancient peoples needed huge amounts of wood for construction, firewood, and shipbuilding. From the first civilizations to Roman times, natural forests in the Mediterranean region, from Lebanon and Greece, Italy and Sicily, to Spain and North Africa, were cleared on a grand scale. This not only led to karst formation and the erosion of entire regions, but would have had an impact on climate as well.

The Industrial Revolution, the invention of the steam engine and the combustion engine, and the mass production of goods marked a turning point in the history of humankind. These developments also revolutionized social and economic structures in previously

GLOBAL SURFACE TEMPERATURES FROM 1850 TO 2009
Deviations from the average temperature for the period 1861 to 2004
in degrees Celsius *

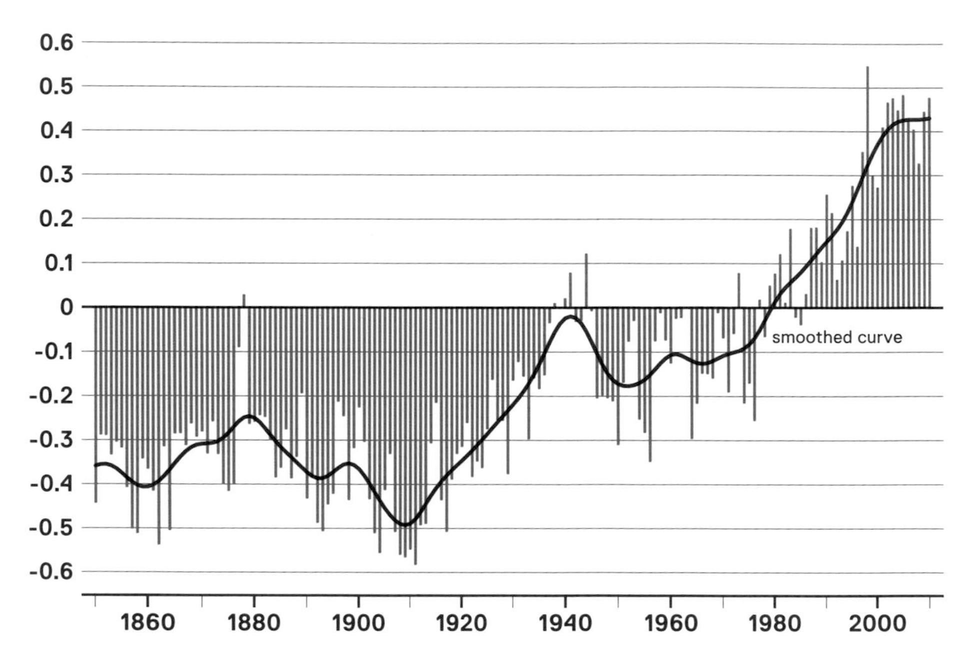

Source: Hadley Centre, 2004, updated by ETH Zurich, 2011
* Note: 1 degree Celsius equals 1.8 degrees Fahrenheit

unimaginable ways and led to a drastic reshaping of the natural environment—and to an exponential increase in the consumption of fossil fuels—coal, oil, and natural gas. In short, the Industrial Revolution turned the human race into a significant climate factor.

When industrialization began in the middle of the nineteenth century, Earth had once again entered a warmer period, which climatologists call the Modern Optimum. Within nearly 160 years, global temperatures rose by nearly 1 degree Celsius (1.8°F), with a rise of 0.7 degrees (1.3°F) in the last century alone. This did not happen continuously because a slightly cooler interval followed the first warming period that ended in 1940. But since 1970, temperatures have been rising almost constantly at the remarkably fast rate of about 0.2 degrees (0.4°F) every ten years. The ten warmest years since the start of systematic temperature registration 150 years ago have all been recorded between 1998 and now.

Natural causes do not sufficiently explain this rapid rise in temperature. A 2007 study by the National Center for Atmospheric

Research (NCAR) concluded that no more than 0.2 degrees (0.4°F) of warming observed in the twentieth century could be attributed to natural causes.

Concentrations of greenhouse gases in the atmosphere have been increasing since the onset of industrialization. CO_2 levels today are more than 40 percent higher than 160 years ago, and the concentration of methane has more than doubled during the same time. All in all, ice cores show that current values are the highest they have ever been in the past 800,000 years.

This increase in atmospheric greenhouse gases is due almost exclusively to the use of fossil fuels, destructive land use like the clearing of rainforests, and other instances of human interference with nature. According to the Global Carbon Project, an unimaginable 1,300 billion tons of CO_2 were released between 1850 and now, more than half of that in the last forty years. Emissions increased by 41 percent between 1990 and 2008.

Since the Industrial Revolution, humanity has altered the global environment to an extent never before seen. Anthropogenic factors such as greenhouse gas emissions, the rate of human population growth and urbanization, air pollution, the large-scale clearing of rainforests, and the building of mega-dams and extensive irrigation systems, further aggravated by the industrialization of agriculture and other forms of drastic interference in ecological systems, have perhaps changed the world's climate to a much greater degree than any natural climate factors. In a seminal article appearing in *Nature* several years ago, atmospheric chemist and Nobel laureate Paul Crutzen therefore suggested that the Holocene had ended and a new epoch had begun—the Anthropocene. This places the responsibility for the development of Earth's climate squarely in our own hands.

If this book were a chronicle of the complete history of Earth, mankind would be mentioned in two sentences at the end, with the last century represented by less than the final dot.

Climate reconstructions prove that today's concentrations of CO_2 in the atmosphere are more than one-third higher than they have ever been in the past 850,000 years.

These cave paintings from the Cederberg region of South Africa bear witness to a considerably warmer climate during the Holocene Climate Optimum from 7,000 to 3,000 BCE. *George Holton/Keystone*

Cave paintings pointing to the existence of fertile river landscapes were discovered in the Tassili n'Ajjer Mountains in the Sahara. They date from the Holocene Climate Optimum. *George Holton/Keystone*

This cave painting of an elephant on the wall in Phillips Cave, Ameib, Namibia, is attributed to prehistoric bush people and dates back at least 4,000 years. *George Holton/Keystone*

Prehistoric hunting scene discovered in a cave in the Tassili n'Ajjer Mountains in southern Algeria. It dates back to a time when water was abundant and the Sahara was lush and fertile. *Mario Fantin/Keystone*

Neo-Assyrian palace on the Euphrates River; the cool, arid climate of the Bronze-Iron Age Cold Period drew people to the banks of large rivers, and early advanced civilizations evolved. *Stapleton Collection/Corbis*

The Carthaginian military commander Hannibal crossed the Alps with 34,000 soldiers, 9,000 horsemen, and 37 elephants of war in October of 218 BCE, taking advantage of the mild climate conditions of the Roman Optimum.

Crop failures and famines during the Migration Period Cooling in the fifth century CE forced Northern European peoples to flee south. *The Sack of Rome by the Vandals,* wood engravings by Heinrich Leutemann. *akg-images*

The Medieval Warm Period facilitated trade, thus contributing to the wealth of cities and their nobility. *Coronation of Baldwin of Flanders as Emperor of Byzantium in 1204,* by Antonio Vassilacchi. *Cameraphoto/akg-images*

The Vikings practiced agriculture and animal husbandry in Greenland during the Medieval Warm Period; by 1550 a renewed climate swing forced them to give up their settlements. Drawing by Rudolf Cronau. *akg-images*

The harsh winters and cool summers of the Little Ice Age led to crop failures and famines. People on the coast of Ireland survived on sea snails and seaweed. *The Illustrated London News*, 1886. *akg-images*

The Little Ice Age brought extremely harsh living conditions to Europe, triggering repeated waves of emigrations. *Emigrants Waiting for Departure to America*, by Felix Schlesinger. *Sotheby's/akg-images*

The spinning machine and the mechanical weaving loom, powered by hydraulic power and steam engines, kicked off the Industrial Revolution in the mid-eighteenth century. 1830. *IAM/World History Archive/akg-images*

Glassblowing was an energy-intensive industry with huge demand for firewood, leading to the demise of extensive forests. *Glassworks at Schliersee,* wood engraving based on a painting by Aloys Eckardt. *akg-images*

Blast furnaces for iron and steel, like this one run by the Siemens Steel Company in Landore, Wales, caused a sharp increase in anthropogenic CO_2 emissions in the nineteenth century. *Oxford Science Archive/Keystone*

A symbol of the twentieth century — humans as victims of the machines they created; Charlie Chaplin in *Modern Times*, 1936. *akg-images*

HOW CLIMATE SCIENTISTS GET THEIR DATA

Today there is considerable certainty about what kind of climate conditions existed in the past. Scientists have gained insight on Earth's climate as it was hundreds of thousands, even millions of years ago. Where does this amazing information come from and how reliable is it?

Klaus Lanz

Before we can make any forecasts about climate trends in the future, we need to know and understand how the Earth's climate evolved in the past. Climate and weather conditions haven't been systematically recorded for very long. The Central England Temperature Series, which began in 1659, is believed to be the longest uninterrupted collection of climate data to the present day. Other, older observations can be found in the meteorological journals kept by William Merle at Oxford from 1337 to 1344. They are an impressive documentation of the very cold weather conditions that prevailed during the Small Ice Age toward the end of the Middle Ages, when even the Baltic Sea froze over twice. We also have records of the Nile River's low and high water levels in Egypt from the seventh century to the present, and indeed, these records extended uninterruptedly from 622 to 1284. Nevertheless, truly reliable information is available on temperature, atmospheric conditions, barometric pressure, and other parameters only for the last 150 years, and data on CO_2 levels only for the past few decades.

WHAT IS PALEOCLIMATOLOGY?

When reconstructing climate history, climatologists take advantage of the fact that the climatic conditions of the past left traces all over our planet. Layer by layer, material settled at the bottom of oceans, lakes, and moors over thousands of years. Scientists today can draw conclusions about environmental and climate conditions at the time of sedimentation by studying the remains of trapped organisms such as mussels, crayfish, plankton, dust, and pollen. The deeper the layer studied, the farther they can look back into the past. Wherever stable conditions prevailed for prolonged periods, scientists can expect to find reliable climate-related data. Old trees are witnesses of past centuries, as corals are of past millennia. Stalactites, stalagmites, and ice from polar glaciers give us insight into the history of the Earth hundreds of thousand of years ago. Sediment layers on the ocean floor are up to 150 million years old, and climate indicators that go even further back can be found in stratified rocks in the Alps and other mountain ranges.

Unlike written records of climate, the information stored in these natural archives is not always easy to read. Plant and animal remains on lake bottoms are relatively easy to interpret. For instance, the pollen contained in varves, sedimentary deposits during a single year, can be identified even thousands of years later. This information directly indicates what type of vegetation existed around a lake, and indirectly indicates what the local climate was like at that particular time. The absolute age of a sample of plant or animal material can be accurately determined based on the decay of radioactive isotopes such as carbon-14.

DENDROCHRONOLOGY: TREES AS HISTORY BOOKS

Tree-ring dating has proven to be a very reliable record of past climate and growth conditions, especially in trees that grow in temperate zones with a marked difference between seasons. In spring, trees produce wood that is lighter in color, called "early wood." Later in the year, before winter sets in and growth stops, they produce a visibly darker wood called "late wood." This creates distinct and countable annual rings. The wider the ring, the more favorable growth conditions were for the tree.

Newly felled trees are not the only keys to climate history. To gather information on periods in the distant past, scientists also examine wood from historical buildings and tree trunks that have been trapped in glaciers for thousands of years. Growth rings have been systematically charted for so many trees that analyzing them has allowed researchers to reconstruct calendars extending far back into the past. The Hohenheim tree ring calendar for Central Europe goes back some 14,600 years. It chronicles periods of favorable and less favorable growth conditions. Yearly growth rings don't tell us whether these conditions were mostly influenced by warmer temperatures or rather by higher CO_2 levels, so to obtain this kind of information, researchers compare this information with other climate archives.

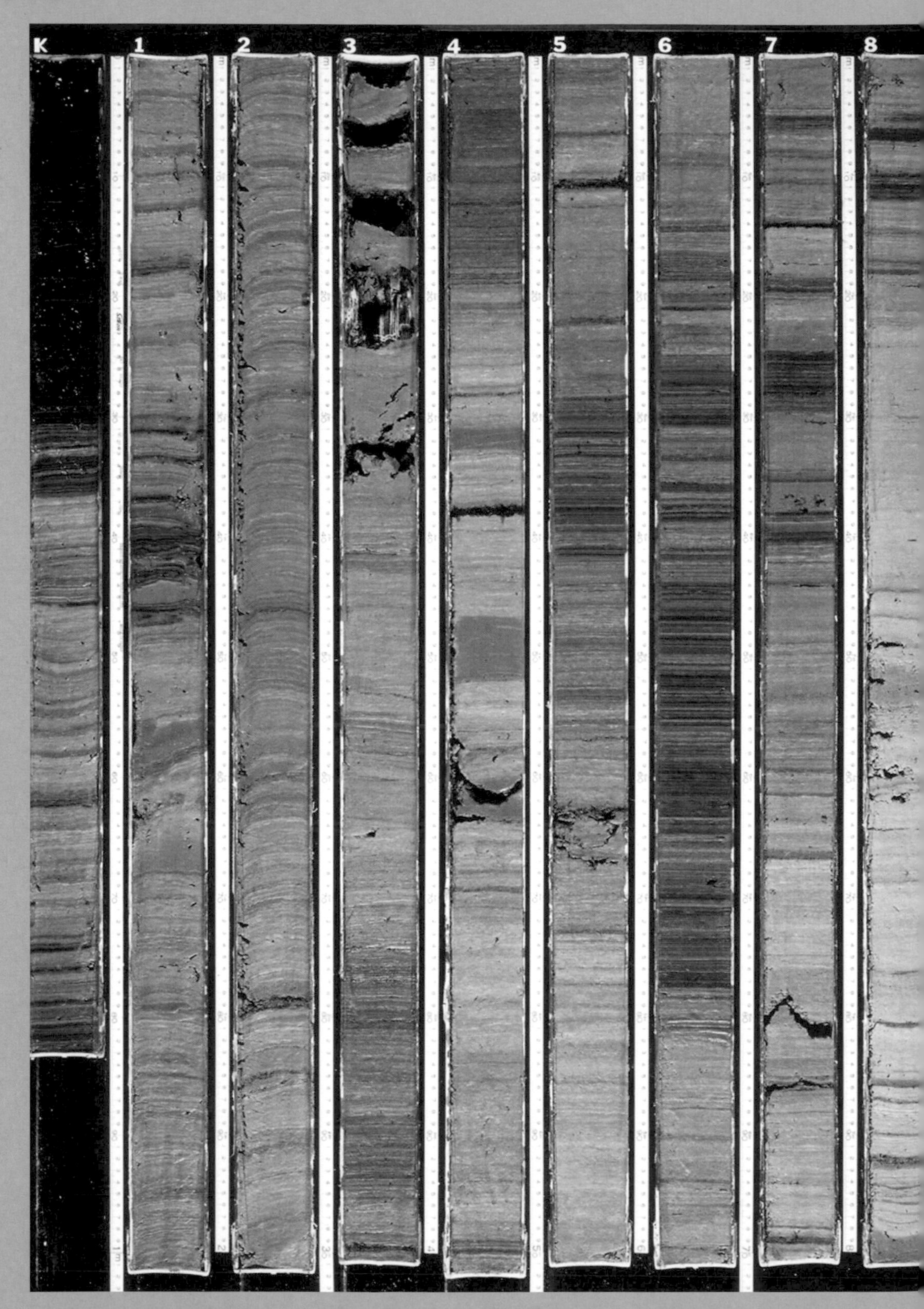

14,000 years of natural history stored in yearly sediment layers (varves) of Lake Baldegg in Switzerland; pollen preserved in these layers gives evidence of the vegetation that once grew around the lake. *Michael Sturm/Eawag*

Water forms dripstones layer by layer. Analyses of the ring-like structures in dripstone cross sections provide accurate data on climate history. *Dominik Fleitmann/University of Berne*

Just as counting the year rings of a felled tree reveals its year of germination, petrified wood found in prehistoric pile dwellings can be precisely dated thousands of years back. *Christof Bigler/ETH Zurich*

ETERNAL ICE

Glacial ice also has a long story to tell wherever layers of ice have been able to build up continuously without melting or suffering any other disruption. Ideal conditions are typically found at the center of the arctic glaciers of Greenland and in Antarctica's ice sheets, which are several kilometers thick. Hundreds of thousands of years of snowfall have accumulated here layer by layer. Deep core drilling is used to extract ice core samples for analysis. It is very rare to find direct plant or animal indicators for earlier climate conditions in these latitudes. But researchers can gain insight into changes in the composition of the Earth's atmosphere by analyzing the bubbles of gas trapped in the ice.

Additional clues to the past can also be found indirectly. Information deduced in this way is referred to as proxy data or climate proxies. For example, researchers can gather information on atmospheric temperatures during earlier periods by analyzing the ratio of isotopes in glacial ice. Climatologists take advantage of the fact that water on Earth can contain three isotopes of oxygen with varying mass. Most H_2O molecules contain the most abundant form of oxygen, the light ^{16}O atom. A few water molecules per thousand contain the heavy ^{18}O oxygen atom, and, finally, there are traces of water with the ^{17}O oxygen atom. Water with the ^{18}O atom is 10 percent heavier than water with the ^{16}O atom and therefore evaporates somewhat more slowly. As a consequence, water vapor is enriched with "light" water, and, inversely, "heavy" water condenses more rapidly.

On its way to the cold, polar latitudes, heavy (^{18}O) water tends to rain or snow down more readily, so that the remaining share of heavy water in water vapor is lower. The farther that water vapor advances into the cold regions, the less heavy water it contains, and, consequently, the less is rained or snowed down. Therefore, the ratio of oxygen isotopes stored in snowflakes gives an indication of the temperature of the atmosphere. These snowflakes, preserved in glacial layers, enable scientists to reconstruct and chronicle temperatures in the past. The less heavy water they find in a glacial layer, the lower the average atmospheric temperature must have been at the time those snowflakes fell. In well-preserved glacial ice, this even makes it possible to differentiate summer rain from winter rain or snowfall so that annual layers can be identified in ice cores.

In other words, the ratio of $^{18}O/^{16}O$ isotopes in ice cores gives researchers extensive information about temperature fluctuations in the area surrounding the respective glacier. In Greenland,

The snow laboratory of the Byrd Station in Antarctica, 1961; researchers are still unaware that the snow and ice beneath them store hundreds of thousands of years of climate history. *Albert Moldvay/Getty Images*

this information goes back 123,000 years, and in Antarctica as far back as 800,000 years. Seawater temperature data can be obtained by analyzing the ratio of $^{18}O/^{16}O$ isotopes in the fossils of microscopic marine organisms, diatoms, and corals found in sediments on the ocean floor. These organisms build varying amounts of both oxygen isotopes into their outer cell walls. The lower the sea temperature, the higher the share of ^{18}O. Using this method, researchers have succeeded in chronicling temperature fluctuations in the oceans as far back as 60 million years ago.

STALACTITES AND STALAGMITES

In search of witnesses of Earth's history that are even more reliable and easier to date, researchers turned to caves containing dripstones. These dripstones, called stalactites and stalagmites, are formed over thousands of years by the dripping of crevice groundwater containing dissolved calcium carbonate and other minerals, including traces of natural uranium. Researchers are able to determine the age of dripstones by making use of the measurably slow rate of radioactive decay of uranium into thorium and lead. Like trees, dripstones have concentric annual rings that grow outward from the center. The rings become visible in a horizontal cross section, allowing researchers to study them.

Uranium dating is very accurate. The age of 50,000-year-old stalagmites from caves in northeastern Turkey could be determined to within 140 years—meaning that the level of uncertainty in dating is well below 1 percent. Using measurements of $^{18}O/^{16}O$ isotope ratios as well, researchers at the University of Berne in Switzerland were able to deduce how warm it was when the original groundwater first evaporated from the ocean and then precipitated as rain. Dripstones provide historical records for regions in which it is too warm for glaciers to be reliable climate archives.

BETTER AND BETTER DATA

Every paleoclimatic method and the proxy data gathered from it carry with them a certain level of uncertainty. Accuracy can be improved by comparing, combining, and consolidating findings from several different dating methods. Today we have information that chronicles the fluctuations of local, regional, and global climate over several hundred million years. However, the farther we look into the past, the more difficult it becomes to interpret proxy data. Then, as now, climate conditions on Earth always varied greatly from place to place. Some proxies describe local or regional conditions while others delineate global changes, for instance in the carbon cycle. We always need to combine local and global data to be able to draw reliable conclusions on the global climate conditions of the past.

It is fascinating how paleoclimatologists can gather information on long-past eons of the Earth and its climate's history. Increasingly accurate analytical data, along with new and more sophisticated methods, combine to play an important role in this research. Civil society needs reliable scenarios for the future to find ways to cope with climate change. The better we learn to understand Earth's history, the more reliable these projections can be.

THE CLIMATE SYSTEM

WHY IT'S GETTING WARMER

Christian Rentsch

TING

R

Almost everyone agrees that global temperatures, notwithstanding considerable fluctuations, have been rising more or less steadily for the past 150 years—and hardly anyone disputes the fact that atmospheric levels of greenhouse gases such as carbon dioxide have increased significantly. But this is where consensus ends, at least as far as the general public and politicians are concerned. There is no societal consensus on what causes climate change, or to what degree, if at all, greenhouse gas emissions generated by human activity are to blame.

Couldn't other, entirely natural factors have caused global warming? After all, haven't climate changes always taken place, even to a larger extent, long before humans ever had a hand in things?

Climatologists have answers to these questions, but they are multifaceted and not always straightforward. The climate system is exceedingly complex, and changes in the system result from many processes that interact in extremely complicated ways and often mutually intensify or diminish each other's effect. Solar activity plays as much a role as changes in the Earth's orbit around the sun, and a number of additional factors associated with the Earth system are equally significant.

For example, the greenhouse effect is far more complicated than the term might lead us to believe. The Earth's atmosphere cannot be compared to a simple greenhouse. Wind systems, clouds, and ocean currents transport and distribute heat and humidity across different climate zones, influencing developments in climate, as do alpine glaciers and the huge ice masses at the planet's poles. The plant and animal world, as well as changes in vegetation brought on by logging operations and new agricultural uses, affect atmospheric levels of carbon dioxide and other greenhouse

gases as much as the global consumption of coal, oil, and natural gas does.

In short, there are no simple answers to these simple questions. Most relationships between the factors affecting climate can't be measured directly, even less so their overall interaction. They become accessible only indirectly by combining empirical data, laboratory experiments, theoretical computations, mathematical climate models, and computer simulations. Climate researchers therefore deliberately choose to use different methods and models to gather information. Working in this way helps scientists identify false assumptions because these can't be confirmed when a full range of models or methods is used. Likewise, research findings that are repeatedly validated using several approaches and different paths have a high level of reliability.

Moreover, climatology is a young science. The systematic worldwide measurement of even the most basic weather data did not begin until late in the eighteenth century. Indeed, only in recent decades have researchers been able to measure more and more data directly—an absurdly short period of time for a scientific endeavor that deals with processes covering thousands and millions of years. Information about climate changes in the distant past has to be indirectly reconstructed with the help of climate archives such as ice cores and sediments; the necessary instruments have been developed only in recent decades. Finally, only the relatively recent development of supercomputers has allowed scientists to compute complex relationships in today's changing climate and to make comparisons with the proxy data on ancient climate history found in natural climate archives.

The Intergovernmental Panel on Climate Change (IPCC) commissions several large working groups to put together information on the most recent findings in climate research, and then publishes the results every few years in its IPCC reports. These reports show there is hardly any doubt today that human intervention in our planet's environment, in particular through the emission of greenhouse gases, has been responsible for global warming since the beginning of the industrial era.

But not even the IPCC reports can predict how climate will change in coming decades or up to the end of the twenty-first century, or beyond that. Climate change will depend on the development of industry, commerce, technology, and the world's population, on how the global community deals with environmental issues, and on whether we continue to rely on fossil fuels for our energy needs. It will also depend on how willing we are to take steps to protect the climate. Climatology can provide us with various scenarios to help us decide what to do. Ultimately, we humans are the ones responsible for the future of our planet's climate.

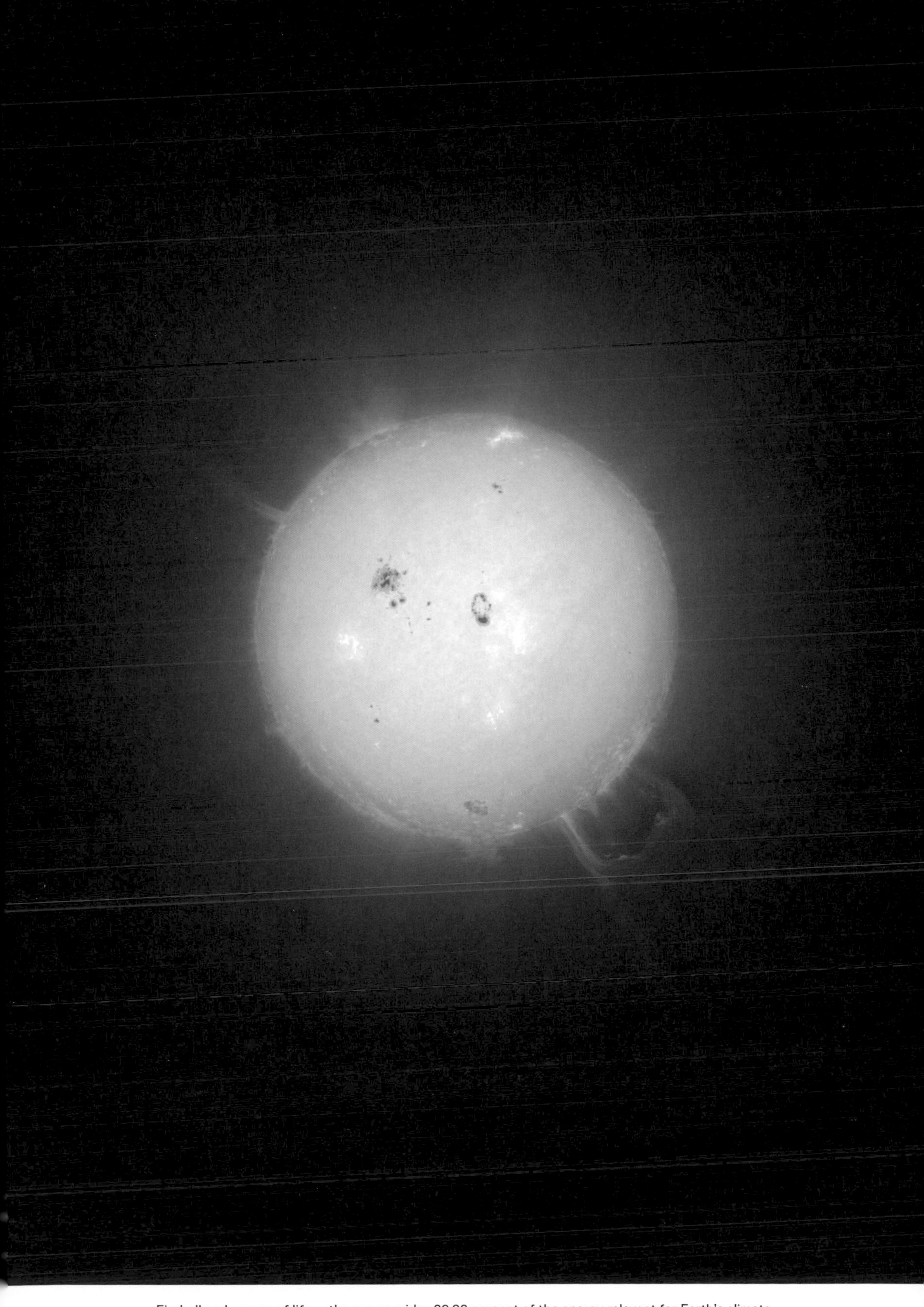

Fireball and source of life — the sun provides 99.98 percent of the energy relevant for Earth's climate.
Detlev van Ravenswaay/Science Photo Library/Keystone

Seen from outer space, Earth's atmosphere is a fine and fragile envelope. It is our planet's weather and climate lab. *NASA/Keystone*

The oceans drive the global water cycle, with some 1,300 billion cubic meters of water evaporating every day.
Zen Shui/Milena Boniek/Getty Images

Crucial for the Earth's radiation balance — ice and snow masses reflect 80 to 90 percent of the solar radiation they receive back into space. Sidujökull Glacier, Iceland. *Arctic Images/Corbis*

Dust particles in the atmosphere govern the formation of clouds and reduce the intensity of solar radiation on the Earth's surface. A sandstorm blows across the Namib Desert. *George Steinmetz/Corbis*

Forests are important CO_2 sinks and have become a key issue in international climate negotiations.
A tropical rainforest in Panama. *Will and Deni McIntyre/Keystone*

Double trouble — fires destroy valuable forests and also contribute to the greenhouse effect. Stafford Township, New Jersey, USA. *Noah Addis/Corbis*

Volcanoes spew enormous amounts of gas and dust into the atmosphere, but they have very little influence on current changes in our climate. Eyjafjallajökull, Iceland, April 2010. *Arctic Images/Corbis*

THE SUN KEEPS US WARM

Sumerians and Egyptians worshiped the sun in cult rituals and chanting, but modern climate scientists use a prosaic number to describe the sun's importance: 1.74 × 108 gigawatts. It is not easy to grasp the enormity of this number—it is 463,000 times the capacity of all the world's nuclear power plants combined. This energy has kept Earth from being a dead planet with an unimaginably cold temperature of about 273 degrees Celsius below zero (-460°F).

The sun provides 99.98 percent of the Earth's energy. The Earth's own internal heat is so minimal in comparison that it plays no role in determining global temperatures.

The flow of energy from the sun to the Earth is not constant, having increased by 30 to 40 percent since the Earth first formed 4.6 billion years ago. It will continue to increase in astronomical periods of time until it reaches its peak in five and a half billion years. But by that time, life on Earth will have long since disappeared—astrophysicists have calculated that in two or three billion years, the sun's energy will have heated the Earth's climate to more than 100 degrees Celsius (212°F). Then, after another few dozen billion years, the sun, having exhausted all its energy, will turn into a white dwarf as it fades in the universe.

Astronomical dimensions like these are not relevant for our climate. More importantly, solar activity also goes through variations in shorter spans of time referred to as sunspot cycles. Sunspots are signs of irregularities in the solar magnetic field. They are easily visible from Earth, but very small in comparison to the sun itself, measuring 1,000 to 50,000 kilometers in diameter. They are about 2,000 degrees Celsius (3,600°F) cooler than the rest of the sun's surface, which is 6,000 degrees Celsius (10,832°F). Approximately every eleven years, the number of sunspots increases for some time, with spots appearing spontaneously, often in groups, and disappearing after a few days to weeks. Their appearance is always accompanied by an increase in electromagnetic radiation.

It still isn't entirely clear whether these eleven-year variations in solar activity have any impact on Earth's climate, or better said, on its surface temperature, and if so, how large the effect is. The fact that the Small Ice Age, which lasted from 1250 to 1850, peaked during an unusually long period of time without sunspots, leads some climate scientists to believe that the influence of sunspots on our planet's climate must be considerable, although they don't know to what degree solar activity was reduced at that time. There is the possibility that a higher number of volcanic eruptions could have been responsible for the drop in temperature. Modern-day measurements show that sunspots cause a variation of only about 0.15 percent in the intensity of solar radiation—actually not enough to directly and markedly affect our climate.

Three other factors related to the sun have a much greater influence on Earth's climate—the shape of the Earth's orbit (eccentricity), the tilt of the Earth's axis (obliquity), and gravitational interactions between Earth, other larger planets, and the moon (precession); taken together, their interactions are referred to as the Milanković cycles →p. 124. They were first calculated in the 1920s by mathematician and astronomer Milutin Milanković.

In astronomy, eccentricity refers to a slight variation in the Earth's orbital path around the sun. It changes from a more circular to a more elliptical form within a 95,000-year cycle. When the orbit becomes more elliptical, year-round solar radiation decreases only slightly, but substantial changes result in terms of how much solar insolation the Earth receives during each season.

Astronomers use the term obliquity to refer to periodic changes in the Earth's axial tilt. If the planet's axis were perpendicular to the orbital path, the Northern and Southern Hemispheres would receive the same regular amount of sunlight all year round and there would be no seasons. The more the axis deviates from the perpendicular, the greater the difference in the amount of sunlight reaching southern and northern latitudes during the seasonal course of the year. The axial tilt vacillates between 22 and 24.5 degrees during a cycle of 41,000 years.

Precession describes a kind of disruption in the Earth's axis of rotation caused by gravitational tidal forces exerted by the moon and larger planets in our solar system. These forces make Earth's axis move in a way best compared to the wobble of a spinning top. Precession causes the seasons to shift relative to the Earth's orbital path. For instance, the occurrence of northern and southern summers

MILANKOVIĆ CYCLES

Eccentricity

The Earth's orbit around the sun varies between a more circular and a more elliptical shape in a cycle lasting 95,000 years.

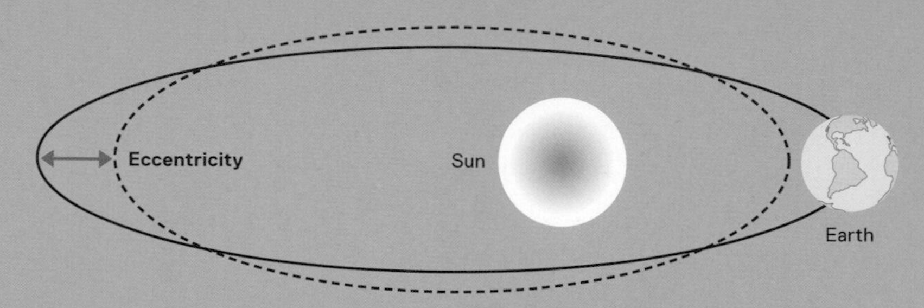

Obliquity

Over the course of 41,000 years, the tilt angle of the Earth's axis varies in relation to the orbital plane.

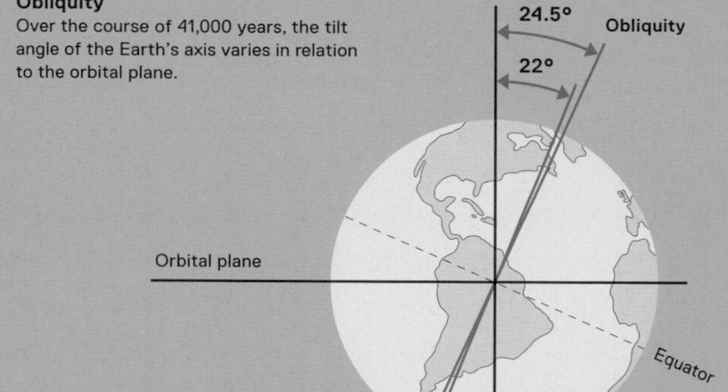

Precession

During a cycle that lasts between 19,000 and 23,000 years, the Earth's axis wobbles around an axis perpendicular to the orbital plane, like a spinning top.

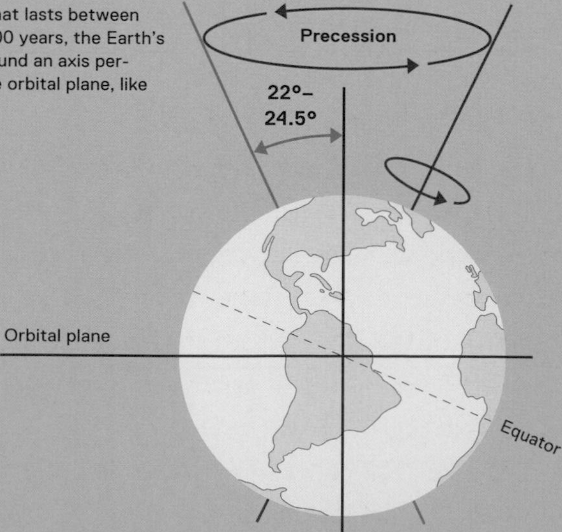

Source: Redrawn from Cuffey & Brook, 2000

slowly shifts from the closest to the farthest point in Earth's orbital path around the sun, following the course of two different cycles, one lasting 19,000 and the other 23,000 years.

The superposition of these three cycles of varying length results in irregular fluctuations in the intensity of solar radiation in Earth's Northern and Southern Hemispheres and in respectively longer and shorter summer and winter seasons. This is particularly noticeable in the higher latitudes of the Northern Hemisphere where large expanses of land warm up faster than the oceans that cover most of the Southern Hemisphere.

Milanković's calculations in the 1920s showed an approximate concurrence with the change from glacial periods to warm periods over the past 3 million years; during the first 1.8 million years, individual cycles lasted roughly 41,000 years, and since then they have lasted about 100,000 years. Glacial periods lasted about 90,000 years and warm periods about 10,000 years.

The fact that Milanković cycles have had an impact on climate changes over millions of years is indisputable, but the question to this day is *how.* On a global average, the Milanković cycles account for fluctuations of only 0.17 percent in total solar radiation. Even if their effects in different latitudes and on seasonal changes were greater, the Milanković cycles on their own, without any further climate feedback to intensify their influence, could not have caused the swings between glacial and warm periods.

The ice-albedo effect, also referred to as ice-albedo feedback, is one such mechanism that boosts the impact of the Milanković cycles. If climatic cooling and longer winters cause more snow to fall which doesn't melt away for longer periods, not even during the summer, the albedo—the direct reflection of sunlight away from the surface of the planet—becomes stronger, cooling the climate even more.

Or if temperature and pressure differences between latitudes are altered as a result of the Milanković cycles or other factors, the intensity of global wind and ocean currents also changes, playing a vital role in climate change.

Climatologist Eli Tziperman suspects that the Milanković cycles function only as a kind of pacemaker nudging other processes in the climate system and "coordinating" their phases (phase-locking).

In this way even weak Milanković "signals" could trigger considerably greater climate changes.

One thing is certain, however—neither changes in solar activity, nor sunspots, nor Milanković cycles are the cause of today's unusually rapid rate of global warming of 0.7 degrees Celsius (1.3°F) over the course of the past one hundred years.

THE GREENHOUSE EFFECT: EARTH'S ENERGY RECYCLING PROGRAM

Our climate is the result of innumerable physical, chemical, and biological processes taking place simultaneously in the atmosphere and oceans, and on the Earth's surface, interacting and responding to each other. If we wish to understand these processes as a single integrated system, we need a common denominator enabling us to quantitatively connect such distinctly different phenomena as thermal radiation, differences of pressure in the atmosphere, and wind intensity. This common denominator is energy.

Energy comes in many different shapes and forms. It can be transformed from light into heat or kinetic energy; it can be chemically stored in a battery and then converted into light by a lightbulb; or oceanic systems can relay energy to atmospheric systems. In short, the planet's climate can be portrayed as a system based on energy flow. Or as climate scientists Stefan Rahmstorf and Hans Joachim Schellnhuber say, "Our climate comes from a simple balance of energy in a global setting."

The concept of a global energy balance is, of course, a very simplified model of reality. It does not give us information on differences in temperature between seasons or various climate zones—all it does is describe the long-term global average temperature. Nonetheless, the concept of an energy balance helps to understand and calculate the climate system as an entity. Basically, if the amount of energy supplied to Earth from space is the same as the amount it releases back out to space, the flux of energy is balanced. If Earth releases less energy back to space than it receives (because of a change in the strength of the greenhouse effect, for example), the climate system loses its balance. Then, the system will respond by adapting its fluxes of energy until the energy balance is restored. In other words, average global temperatures will keep increasing until Earth releases as much energy as it receives.

Our climate system gets 99.98 percent of its energy from the sun. Solar energy is made up of electromagnetic radiation of very different wavelengths, the spectrum ranging from X-rays and ultraviolet light waves to visible light waves to long, warm infrared radiation. The most intensive rays of the sun are visible light. “Light energy” makes up roughly half of solar energy.

Not all rays share the same fate on their way through the atmosphere and on the Earth’s surface. Most ultraviolet light is absorbed by the ozone layer in the stratosphere, and only a small part of solar infrared radiation reaches the Earth’s surface, whereas most (visible) light energy penetrates the atmosphere unhindered.

But how does solar energy turn into what we commonly refer to as heat, which is, to be more precise, thermal energy or infrared radiation? Solar energy is transformed into heat when electromagnetic radiation reaches materials; for example, when sunlight hits the Earth’s surface, water, or even our skin. Depending on their properties, these materials absorb a certain amount of sunlight, convert it into thermal energy, and re-emit it. This re-emitted thermal energy determines temperatures on the surface of the Earth, the water, or our skin.

The fluxes of energy through the climate system can show us how different aspects of climate are linked and influence one another, especially in terms of one key factor—global warming.

The starting point is the energy reaching Earth from the sun; in other words, the input to the energy balance →p.129. Climate scientists refer to the long-term average amount of energy arriving on the Earth from the sun as the solar constant. This is the same as the 1.74 × 108 gigawatts mentioned earlier →p.122. Climate physicists use a different figure derived from that value; based on Earth’s total surface area, the incoming radiation from the sun amounts to 342 watts per square meter. This is the average amount of energy that would reach every square meter of the Earth’s surface if the Earth did not have an atmosphere.

But the flux of solar energy gets reduced on its way through the atmosphere. Some of the sun’s energy is reflected immediately back into space. How large this share is, called atmospheric albedo, depends on the transparency of the atmosphere. Stratus clouds reflect 40 to

60 percent of the incoming energy, cumulus clouds as much as 70 to 90 percent, and even aerosol particles reflect some of the sun's radiation. The atmospheric albedo is on average about 23 percent.

Another share of the solar energy flux, again roughly 23 percent, is absorbed by the atmosphere. This includes most of the ultraviolet radiation and almost all of solar infrared radiation. Unlike the solar radiation reflected back into space, this share stays within the climate system and has a thermal effect.

The remaining solar energy fluxes, about 54 percent of the original (or 184 watts per square meter), reach the surface of the planet. Climate scientists call this share "downward solar flux" (or even more specifically, the "All-Sky Surface Downward Shortwave Flux").

A fraction of this downward solar flux is reflected back to space from land and water surfaces, but the greater share is absorbed and converted into thermal energy. How much is reflected back depends on the nature of the reflecting surface. Water surfaces reflect back between 7 and 25 percent, while snow and glaciers reflect between 30 and 95 percent, with the fraction depending heavily on the incident angle of solar radiation. Ground surfaces reflect solar radiation to greatly varying degrees, which is particularly significant for regional climates. Desert areas reflect between 20 and 40 percent of the sun's radiation, and grasslands, grain fields, evergreen forests, and deciduous woodlands only between 5 and 25 percent. Altogether, the total terrestrial albedo is only about 7 percent (or 23 watts per square meter).

Why do we need these figures? They provide information on whether Earth's energy balance is in equilibrium or not. If albedo changes in response to changes in cloud cover or the concentration of aerosols, for instance, or when glaciers melt, deserts expand, and rainforests are logged, then the share of the solar energy absorbed by the atmosphere and Earth's surface is also altered. The absorbed fraction is converted into heat and directly influences temperatures on Earth.

Let's look at these numbers again. Of 342 watts per square meter of total incoming solar radiation, 30 percent (or 102 watts per square meter) is reflected back to space. The remaining 70 percent (or 239 watts per square meter) is absorbed by the atmosphere and Earth's surface, meaning that most of it is transformed into thermal energy.

ENERGY BALANCE

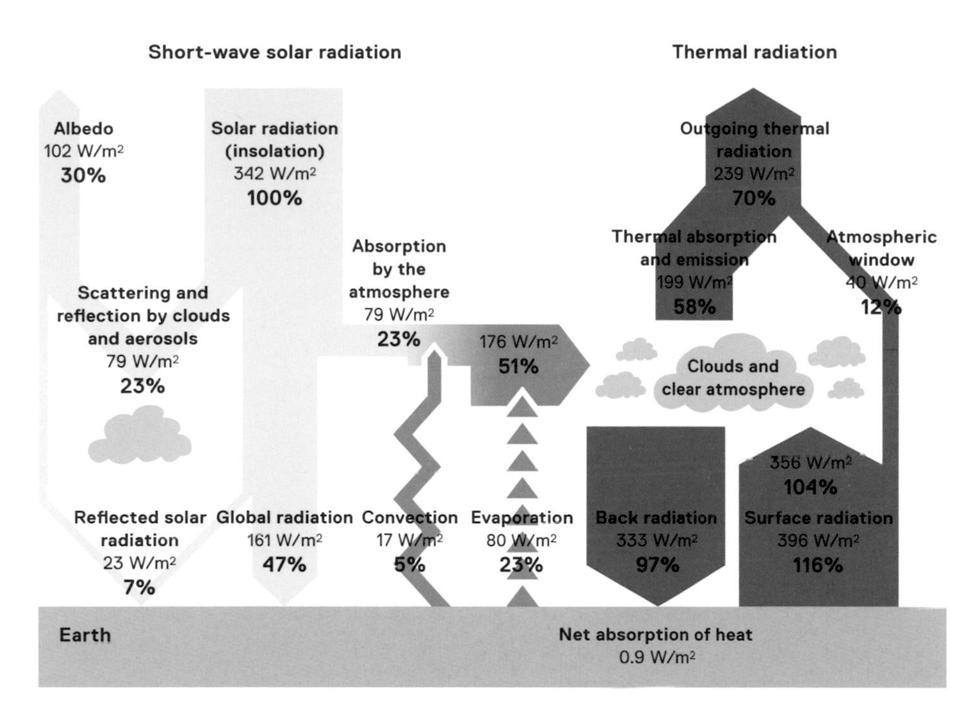

Source: Adapted from Trenberth et al., BAMS, 2009

Using the Stefan-Boltzmann law, a physical formula worked out by physicists Josef Stefan and Ludwig Boltzmann, we can calculate how warm the surface of an object becomes when it is exposed to a certain amount of energy. According to this law, solar energy on its own is enough to heat the Earth by a massive 255 degrees Celsius—up from minus 273 degrees (-460°F), the "absolute zero" temperature of outer space, to minus 18 degrees (0°F).

But Earth's average temperature is currently about 15 degrees Celsius (59°F). This 33-degree difference (59°F), which makes Earth a livable planet, is due to the planet's natural greenhouse effect. To put it simply, the greenhouse effect is a mechanism which increases the efficiency of solar energy and works like an energy recycling system, preventing some of the thermal energy emitted by the planet's surface from escaping into outer space. Instead, greenhouse gases trap this heat in the atmosphere and re-radiate some of it back to Earth. This feedback has caused the average surface temperature on Earth to rise by the aforementioned 33 degrees.

The greenhouse effect—both natural and man-made—is due to the special properties of some gases, the greenhouse gases. If molecules of these gases are hit by thermal radiation, they are activated and begin to vibrate and rotate. They absorb thermal energy and convert it into kinetic energy, which is transformed back into thermal energy immediately or shortly afterward; this heat is released back into the surroundings. Greenhouse gases include water vapor (water in its gaseous form), carbon dioxide, methane, ozone, nitrous oxide (laughing gas), and several other trace gases such as halogenated hydrocarbons.

Greenhouse gases allow most of the incoming solar short-wave energy from the sun to pass through the Earth's atmosphere unhindered, but trap and absorb the thermal (long-wave) energy released back out to space from terrestrial and water surfaces. Greenhouse gases then evenly emit in all directions the thermal energy they have absorbed—much like a radiator gives off heat. Although they get their thermal heat mostly from below, greenhouse gases radiate it in all directions, so that only half goes toward outer space. The other half is radiated back to Earth. The level of this back radiation depends, among other things, on the concentration of greenhouse gases in the atmosphere; back radiation currently stands at about 333 watts per square meter.

Thanks to the greenhouse effect and its recycling of thermal energy, Earth's average global surface temperature is plus 15 degrees Celsius (59°F) instead of minus 18 degrees (0°F). But humans have begun to interfere with this recycling system. The composition of the atmosphere has been altered due to man-made greenhouse gas emissions. An incremental difference, an imbalance of 0.9 watts per square meter, now exists between the amount of energy absorbed and the amount of energy reflected by Earth. This imbalance is responsible for the current rate of global warming.

Water vapor plays a special role in the greenhouse effect. The level of water vapor in the air depends on Earth's temperature; this does not apply to other greenhouse gases. The higher the temperature, the more water evaporates from ground and water surfaces, and the more water can be stored in the air, so that the level of water vapor, responsible for roughly 60 percent of the greenhouse effect, increases or decreases depending on the temperature. If Earth's temperature rises because of an increase in concentrations of other greenhouse gases, water vapor levels in the air also rise, further intensifying the greenhouse effect.

Water vapor transports latent heat, as climate scientists call it, from Earth's surface to the atmosphere. This is a consequence of the fact that when water evaporates to form water vapor, it absorbs heat from the surroundings. This energy is released again when water vapor condenses to form liquid water drops. Through this process, heat can be transported over hundreds of kilometers.

The transport of heat from the surface to the atmosphere occurs in different ways. Most of the heat flux from the Earth's surface occurs in the form of infrared radiation; another share rises as latent heat into the atmosphere with water vapor, while the heat that we sense best, i.e., the warm air that rises upward, makes up only the smallest part.

Another look at the energy balance shows us that the thermal fluxes (199 watts per square meter) radiated out to space by greenhouse gases, together with albedo (102 watts per square meter) and a small share of the energy fluxes (40 watts per square meter) emitted by the Earth's surface in wavelengths that greenhouse gases allow to pass, amount altogether to 341 watts per square meter. This current total output of energy by the Earth contrasts with the energy input from the sun of 342 watts per square meter. In other words, this difference of roughly 1 watt per square meter between the amount of incoming and outgoing energy causes the average global temperature to continue to rise until the energy balance is restored.

The greenhouse effect is a completely natural physical process. Without it, life on Earth would not be possible. It can be simulated in laboratory experiments and it is accepted as a scientific fact. The focus of today's climate debate is not on the natural greenhouse effect, but rather on the impact of increasing emissions of greenhouse gases generated by human activity on global warming.

“Climate is an angry beast and we’re poking it with sticks.”

 Wallace Broecker, oceanographer and climate scientist

THE CLIMATE SYSTEM: A FACTORY TOUR

Even those familiar with climate science can quickly lose sense of how our climate functions. For the moment, let us pretend the climate system is a factory that produces climate. If we went on a tour and visited different departments, we would get the following picture:

The *atmosphere* is the actual weather and climate laboratory. Clouds and aerosols regulate the amount of solar radiation that reaches Earth's surface. They reflect about one-fourth of incoming sunlight back to outer space immediately. This atmospheric albedo is responsible for keeping temperatures cooler on Earth. At the same time, clouds and atmospheric pollution also contribute to warmer temperatures. But it is the greenhouse gases that are the major agents in keeping the planet warm; they are in charge of making sure that a large part of the thermal heat emitted from Earth stays in the atmosphere. Wind systems are responsible for the dispersion of heat, rain, and snow around the world; without them, continents would be barren rock landscapes.

The *oceans* produce water vapor, the "material" of clouds. More than 500,000 cubic kilometers of water evaporate from the oceans every year; this is 500,000 billion cubic meters of water, seven times the amount that evaporates from land areas. Although more than 90 percent is returned directly to the oceans in the form of rain or snow, wind systems distribute at least 44,000 cubic kilometers of water to continental land masses. Ocean currents transport large amounts of heat from the tropics to colder northern and southern latitudes; they are responsible for adjusting temperatures between various latitudes and also between the Atlantic, Pacific, and Indian oceans. At the same time, the oceans act as a kind of heat buffer. Because they warm up more slowly than land areas, they slow down the increase in the Earth's temperature. Oceans also absorb from the atmosphere about 30 percent of annual CO_2 emissions, thereby decelerating the increase in the levels of this important greenhouse gas.

Land surfaces and the oceans absorb most incoming solar energy and then release it back to the atmosphere in the form of heat—to what extent depends on the nature of the ground surface and vegetation. In conjunction with the biosphere, they generate the major share of the greenhouse gases carbon dioxide, methane, and nitrous oxide, but, conversely, the biosphere also consumes carbon dioxide. Like oceans, land areas also produce water vapor. Some 71,000 cubic kilometers of water evaporate each year from these areas. In short, oceans, land surfaces, and the biosphere collectively regulate greenhouse gas concentrations in the atmosphere. Different kinds of landscapes, mountain ranges, and plateaus steer local wind systems across the continents. Topography is also a determining factor for the location of mainland glaciers and winter snows.

The *cryosphere,* made up of the mainland glaciers of Antarctica and Greenland, sea ice, alpine glaciers, and winter snow surfaces, reflects 80 to 90 percent of the incoming solar energy flux back to outer space. Together with water surfaces, the atmosphere, the biosphere, and the lithosphere, i.e., the solid outer crust of the planet, it regulates the albedo. Finally, glacier growth and glacier melt influence sea levels and ocean currents.

As we can see, the climate system is a highly complex enterprise which links a vast number of individual processes at global, regional, and local levels.

Feedback processes are among the climate system's most complex mechanisms. They intensify some initial climate perturbations (positive feedback), and weaken others (negative feedback).

Several times in the course of climate history, one of these positive feedback processes, the snow-albedo or ice-albedo feedback, caused Earth to cool down faster than it would have done if only a diminished greenhouse effect had played a role. When Earth's average temperature drops, the planet has longer winters and cooler summers, with more snow in winter. Snow stays on the ground longer in spring and may not even melt during the summer. The resulting large expanses of snow reflect more solar energy back into space, causing the albedo to increase, and accelerating the cooling of the planet.

But a positive feedback can also accelerate an initial perturbation in the other direction; i.e., it can intensify global warming. For example, in response to an initial increase in greenhouse gases, temperatures rise, causing more water to evaporate from ground and water surfaces. Higher concentrations of water vapor in the atmosphere mean that more thermal energy is absorbed; this boosts the greenhouse effect, in turn making temperatures rise even faster.

But this accelerating dynamic may also give rise to a negative feedback to a certain extent. If more evaporation creates more clouds, low-lying clouds increase the albedo and reflect back more sunlight, resulting in cooler temperatures. On the other hand, high clouds absorb more thermal radiation. Together, these two different feedback processes work like a thermostat, with positive and negative feedback mechanisms regulating each other.

As complicated as this might sound, the reality is even more entangled because feedback mechanisms are also influenced by a number of other factors. To date, the complex workings within the cloud system itself have yet to be fully understood. It is still unclear exactly how clouds form, what factors play a role in the forming of water droplets, in condensation, and for rain or snowfall. We also don't know exactly how high clouds, contributing to warmer temperatures, and low clouds, contributing to cooler temperatures, interact with each other.

The role played by aerosols is not well understood either. Aerosols are tiny solid or liquid particles floating in the air and made of dust, condensed gases, soot or salt particles, and even pollen or bacteria. Hundreds of thousands of these particles are in every cubic centimeter of air (and become visible as smog when the count goes up to a million). Aerosols have an important function in the formation of clouds, and they influence cloud brightness and the condensation of raindrops.

And finally, it is almost impossible to factor in processes that depend on the way humans interfere with nature, for instance with deforestation, changes in landscapes and vegetation due to irrigation, melioration, and the cultivation of water-intensive crops.

Ultimately, there is no way of describing with precision the enormous complexity of the climate system. Even the most powerful computers cannot provide us with accurate simulations or projections, especially because the climate system is a "non-linear, deterministic chaotic system." This means that even though each individual process is governed by physical and chemical principles and therefore subject to strict laws of nature, we cannot accurately compute the

Earth system in its entirety because it is deterministic and chaotic at the same time. More simply said, in reality there are never two fully identical initial conditions. According to the chaos theory, a slight deviation in an air eddy or cloud formation can induce completely differing feedback mechanisms within a very short time that intensify and lead to completely different results (much like a chain of decisions with increasingly serious consequences). Meteorologist and mathematician Edward Lorenz came up with a poetic metaphor for this phenomenon and called it the "Butterfly Effect," coining the term in a talk he gave in 1972 on the predictability of certain climatic events. In his lecture he proved that theoretically "the flap of a butterfly's wings in Brazil could set off a tornado in Texas."

This doesn't mean that climate scientists are unable to provide us with reliable predictions on the climate system. Although meteorologists can't tell us if it is going to rain one year from today, they can, with the help of statistics, rather accurately predict how many rainy days we can expect in the coming year. The same applies to climate predictions. Statistical averages compiled over longer periods of time give us a very accurate idea of the climate system at the global level, and increasingly at the regional level as well. These statistics help us understand how the system works and how it can be expected to develop.

WATER AND WIND KEEP THE CLIMATE SYSTEM IN MOTION

Our Earth is the only planet in the solar system where water is simultaneously present in gaseous, liquid, and solid state. There would be no life on Earth without these three different states of water, and what we normally refer to as climate would not exist either. Water connects oceans, land masses, and the atmosphere in a powerful global transport system that reaches into the farthest corners of the continents and the deepest parts of the ocean.

In this way, the ocean's thermohaline circulation, i.e., the ocean current system that connects all the oceans, makes climatic conditions in many regions north and south of the tropics far warmer and more hospitable than if these regions had to rely on sunlight alone.

The thermohaline circulation is driven by what could be described as natural pumps. In the Norwegian Sea between Greenland and Norway, the warm North Atlantic Current (the continuation of the Gulf Stream northeast) sinks deep down in a powerful current, its suction effect acting like an enormous oceanic pump. The physical principle driving this pump is relatively simple. Cold, salty water is denser and therefore heavier than warm, less salty water, so it sinks to the bottom. This warm North Atlantic Current is relatively salty because the surface waters feeding this current from the Caribbean have experienced strong evaporation, causing the remaining water to become saltier. When these waters reach higher latitudes off the coast of Northern Europe, they cool down, sink by more than 3,000 meters, and flow south as a cold, deep-sea current until they approach Antarctica, where the Antarctic Circumpolar Current channels them into the Pacific and Indian Oceans. A second deep-water formation area is in the Weddell Sea off the coast of Antarctica. In this case it is not evaporation but the formation of sea ice that makes the water saltier and therefore prone to act as a pump.

The warm North Atlantic Current, driven mainly by tropical trade winds, transports up to 150 million cubic meters of water per second and is responsible for the fact that the climate in Western Europe is several degrees Celsius warmer on average than in most other regions at these latitudes. Several times in recent climate history, the massive melting of glaciers in Canada diminished and may even have shut down the inflow of warm water from the South into the North Atlantic. This brought cold periods with severe winters and devastating crop failures to Europe, each lasting from several decades to several centuries. Although climate scientists believe that such an event is highly unlikely within the next few centuries, this discovery captured the imagination of film director Roland Emmerich (*The Day After Tomorrow*), and drew the attention of scientists to volatile phenomena known as tipping elements. These are mechanisms that, if pushed to a tipping point, will cause climate to change dramatically within a very short time. If the thermohaline circulation were to "merely" weaken, or if the melting of Greenland glaciers pushed it farther south, consequences for Europe's climate would indeed be extensive.

Analogous to ocean currents, wind systems are large-scale planetary circulation systems affecting global climate, although their main impact is on regional and local conditions. Among these are the Hadley cells in the tropics →p.139. The Hadley cells are circulation loops extending north and south of the meteorological equator. They are driven by warm air generated in a band around the equator known as the Intertropical Convergence Zone, where very warm air rises all year round. This air, containing a lot of water vapor when it rises over the oceans, moves upward as high as eighteen kilometers into the atmosphere before flowing north and south and sinking back in the subtropics, where it turns into the warm, dry trade winds that blow toward the equator all year long with great constancy. When the humid air rises, most of the water vapor condenses, bringing a lot of rain to the tropical regions that are characterized by this strong lift of air. This is why the world's great rainforests in the Amazon Basin, the Congo area of Africa, and Indonesia are all in this zone. On the other hand, the air sinking in the subtropics is responsible for warm and arid conditions in subtropical regions. This is where deserts and drought-plagued areas lie—the Sahara, the Arabian Peninsula, the arid zones of the American Southwest and Mexico, and the Australian deserts.

Some of the tropical air masses also continue flowing toward the North and South Poles. The Earth's rotation causes these winds to turn steadily toward the east until they flow almost parallel to the latitudes. Jet streams are an example of this kind of air current. They flow in the higher layers of the atmosphere with wind speeds of up to 500 kilometers per hour. This westerly wind flow keeps warm, subtropical air from reaching the polar regions where winds flows in exactly the opposite direction because the Earth's rotation has very little effect near the poles. More or less strong turbulence with alternating warm and cold air fronts in the frontal zones between the temperate climate zones and the polar regions brings unsettled weather to the middle latitudes, while wind circulation in the polar regions is predominantly cold and isolated.

The Indian summer monsoon is one of the most striking climate phenomena. The monsoon enables intense agricultural cultivation with exceptionally high yields in extensive parts of the Indian subcontinent and Southeast Asia but it also causes severe flooding almost every

GLOBAL WIND SYSTEMS BETWEEN THE EQUATOR AND THE POLES

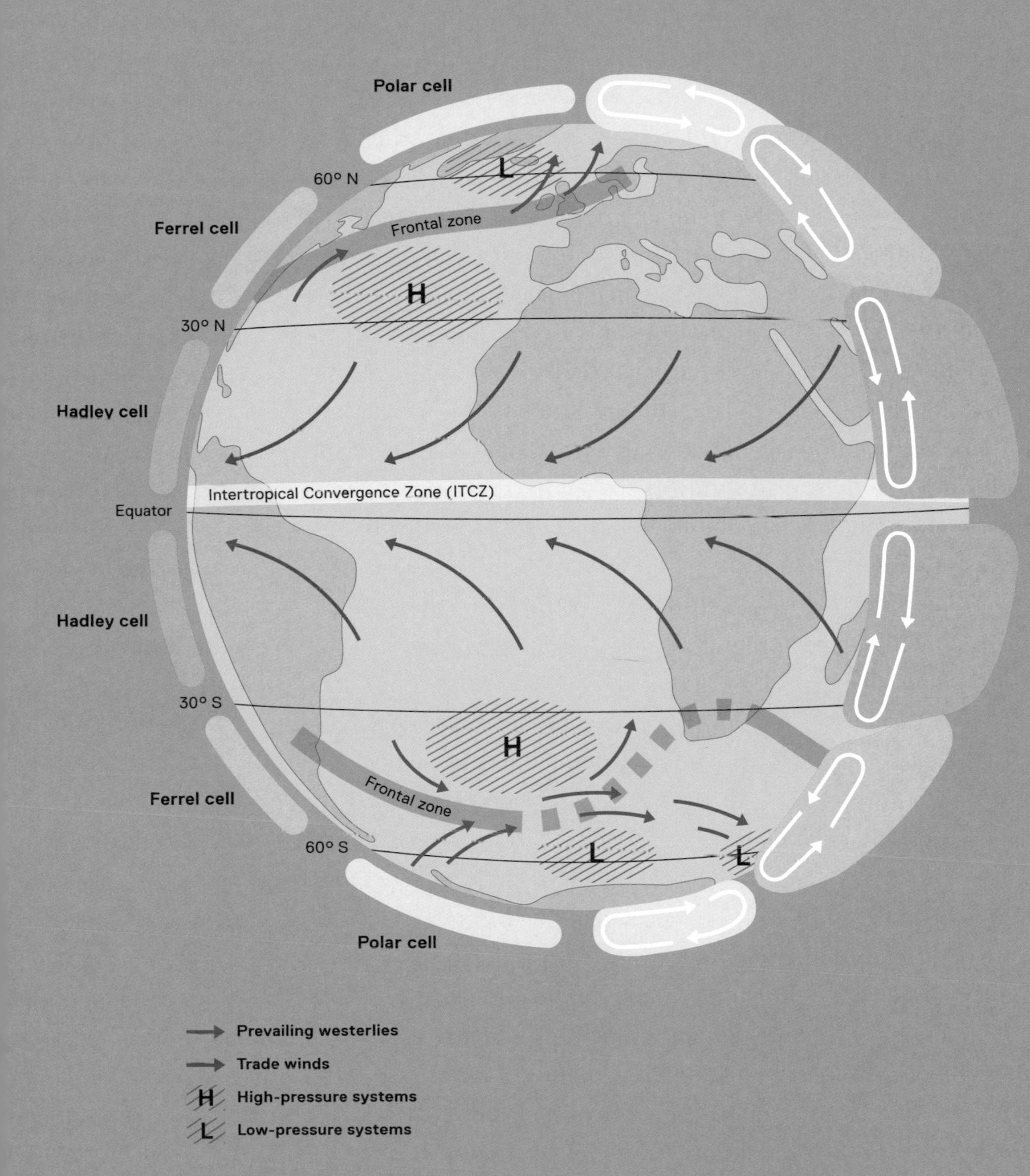

Prevailing westerlies

Trade winds

H High-pressure systems

L Low-pressure systems

Sources: D. Klaus, Die planetarische Zirkulation-Praxis Geografie, no. 6, 1989; NOAA/GOES

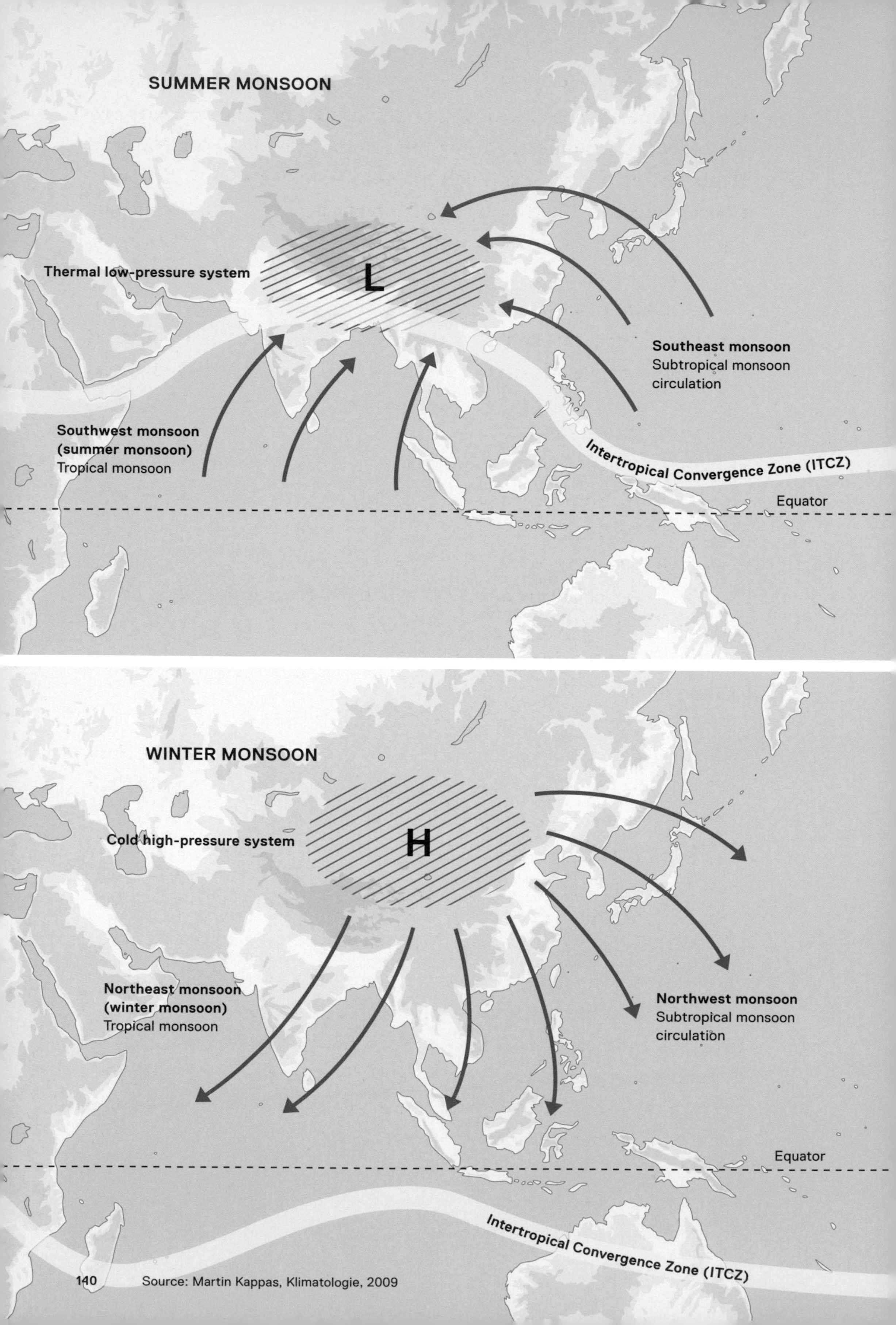
SUMMER MONSOON
Thermal low-pressure system
L
Southeast monsoon
Subtropical monsoon
circulation
Southwest monsoon
(summer monsoon)
Tropical monsoon
Intertropical Convergence Zone (ITCZ)
Equator
WINTER MONSOON
Cold high-pressure system
H
Northeast monsoon
(winter monsoon)
Tropical monsoon
Northwest monsoon
Subtropical monsoon
circulation
Equator
Intertropical Convergence Zone (ITCZ)

Source: Martin Kappas, Klimatologie, 2009

year in India and Bangladesh →p.140. The Indian summer monsoon is the outcome of special weather conditions that occur every year over the Indian subcontinent. During the summer season in the Northern Hemisphere, the sun reaches its zenith at the northern Tropic of Cancer and the Intertropical Convergence Zone shifts northward. Then northeast trade winds over the Indian ocean are replaced by southwest winds that are part of the southern Hadley cell.

As climate scientists now know, the summer monsoon is triggered by intense heating over northwest India and the Tibetan Plateau. Huge masses of rising air form a strong thermal low-pressure region near the surface that sucks in hot air from trade wind zones. Instead of blowing from the Indian mainland toward the equator like trade winds normally do, summer monsoon winds blow in the opposite direction from the equator over the Indian Ocean toward India and Tibet. On their way across the warm Indian Ocean, they absorb large amounts of moisture, bringing torrential rains to the mainland. Rainfall is especially heavy from June to October in the southern foothills of the Himalayas and the catchment areas of the Ganges and Brahmaputra rivers, with some areas receiving up to ten meters of rain.

During the winter, when the thermal low over India and Tibet dissipates and trade wind zones shift toward the Southern Hemisphere, monsoon winds blow back in the opposite direction. The winter monsoon brings cold, dry air from the Tibetan Plateau to India, Bangladesh, and large parts of Southeast Asia. Monsoon systems also exist in western Africa, Central and South America, and northern Australia, but they are not as strong.

Climate scientists still don't know exactly which other phenomena influence the summer monsoon and to what extent. There are connections apparently between the summer monsoon and the ENSO phenomenon, i.e. the El Niño Southern Oscillation. Although there are many uncertainties, climate scientists do agree that global warming will most probably lead to increased monsoon rains and to a higher frequency of extreme monsoon years.

The ENSO phenomenon affects ocean currents and weather conditions in the entire Pacific, if not the entire globe. Peruvian fishermen have known this pattern for centuries and gave it the name El Niño ("the little boy," a reference to the Christ Child), since it occurs for several weeks around Christmas, when it is not worth going out to sea because fish fail to appear. It was not until the 1960s that meteorologist Jacob Bjerknes and other climate scientists realized that El Niño was not just a local event, but actually linked to a large-scale shift in precipitation patterns in the entire Pacific region, even influencing atmospheric circulation as far away as Africa and North America.

In normal years, strong trade winds under the influence of Earth's rotation push surface waters away from the Equator, thereby pulling up cold, nutrient-rich waters from below to the surface. This effect is strongest in the eastern part of the Pacific Basin, and leads to the formation of an equatorial cold-water tongue. Trade winds also push warm surface waters westward along the Equator toward Southeast Asia, where they pile up and create a warm pool. Here, high sea-surface temperatures support strong evaporation and powerful rising motions in the atmosphere, leading to intense rainfall. In contrast, the air above the cold tongue in the eastern tropical Pacific subsides, creating dry conditions over this part of the Pacific.

But in El Niño years, for still not entirely clear reasons, trade winds diminish and consequently the masses of water piled up in the western Pacific begin to flow back to the eastern Pacific. This inhibits the upwelling of cold waters from below, thereby further lowering the difference in temperature between the western and eastern Pacific, which further diminishes the trade winds. This has far-reaching consequences. In El Niño years, hot, moisture-laden air brings rainfall to the mid-Pacific and to otherwise dry South American coastal regions, frequently leading to floods, landslides, and crop failures. In turn, the Western Pacific, Southeast Asia, and Australia experience extreme dryness and drought conditions, which sometimes trigger extensive forest fires in Malaysia and Indonesia.

We still do not know enough about the long-distance effects, called teleconnections, evidently set off by El Niño or the ENSO phenomenon and affecting distant regions of the world. These include heavy rainfall on the west coasts of South and North America, extended dry spells in the Amazon rainforest, a temporarily cooler climate with winter storms and heavy snowfalls on the North American east coast, hurricanes and floods in Mexico, drought, crop failures, forest and bush fires in India and Southeast Asia, torrential rainfall in northern parts of eastern Africa, and increased drought conditions in the Sahel region and southern Africa. Even increased precipitation and winter storms in Europe seem to be linked to the El Niño phenomenon.

El Niño events do not always proceed the same way; they exhibit great variation. For instance, effects on the South American coast and Southeast Asia depend on where the highest water temperatures during an El Niño year occur, in the eastern Pacific or the mid-Pacific. Moreover, oceanographers have discovered that El Niño ocean currents have changed in recent years. It is still uncertain however whether this new kind of El Niño, known as El Niño Modoki, is only a temporary aberration or a potential consequence of global warming.

Teleconnections as a phenomenon pose a number of unsolved problems for climate scientists because they link irregular climatic anomalies occurring in different parts of the globe in very complex ways. Depending on their intensity and the place and time of their appearance, they disrupt "normal" patterns of climatic events. Teleconnections evidently increase the variety or variability of the climate in faraway regions as well as on a global scale. They make it even more complicated for researchers to project how regional climate changes will affect our planet's climate system in its entirety.

HOW GREENHOUSE GASES FUEL GLOBAL TEMPERATURES

In the past one hundred years, the average global temperature has gone up by 0.7 degrees Celsius (1.3°F) and it continues to rise, currently by about 0.2 degrees (0.4°F) every ten years. The ten warmest years since reliable measurement began have all been between 1998 and 2010.

Levels of greenhouse gases in the atmosphere have also risen since preindustrial times. Carbon dioxide (CO_2) has gone up roughly 43 percent, methane (CH_4) by almost 170 percent, and nitrous oxide (N_2O) by about 18 percent.

Finally, anthropogenic CO_2 emissions, including those caused by deforestation and land-use changes have increased from nearly

200 million tons to 34.8 gigatons per year since industrialization began. Some climate scientists estimate that since 1850 approximately 1,700 gigatons of CO_2 attributable to human activity have been released to the atmosphere.

As impressive as these facts may be, they don't prove that anthropogenic greenhouse emissions are responsible for current global warming. Couldn't entirely other factors be the primary cause? Or couldn't carbon dioxide from natural instead of anthropogenic sources have driven the greenhouse effect over the past 150 years?

The climate system is indeed determined by too many intervening factors and too many different kinds of intensifying or diminishing feedback processes for a few data series with similar findings to serve as irrefutable evidence. Proving the connection between greenhouse gases and global warming is very much like building a lawsuit on circumstantial evidence—there is no simple, obvious chain of proof.

To obtain more insight, climate scientists are developing theoretical models based on physical laws and laboratory experiments and check them against empirical findings. Or they compare current climate trends with past climate developments to detect similarities and differences. They test known correlations under varying conditions, using different statistical methods to find out how far they are universally valid. Applying a reverse process, they also attempt to exclude climate factors that may have played a key role in the course of climate history but that no longer explain today's climate trends.

Three such climate factors are solar activity, the Milanković cycles, and the sunspot cycles →p.122. They play an important role in climate history, but don't explain the rapid global warming of 0.7 degrees Celsius (1.3°F) within a single century—average solar activity did not change much during this time. The Milanković cycles need much longer to have an effect on global warming, and the sunspot cycles, which last an average of eleven years, are too short and not strong enough. In any case, temperature developments over the last 250 years do not show any significant concurrence between changes in temperature and sunspot cycles. In the end, only greenhouse gases and their influence on the greenhouse effect can satisfactorily explain the rapid rise in temperature.

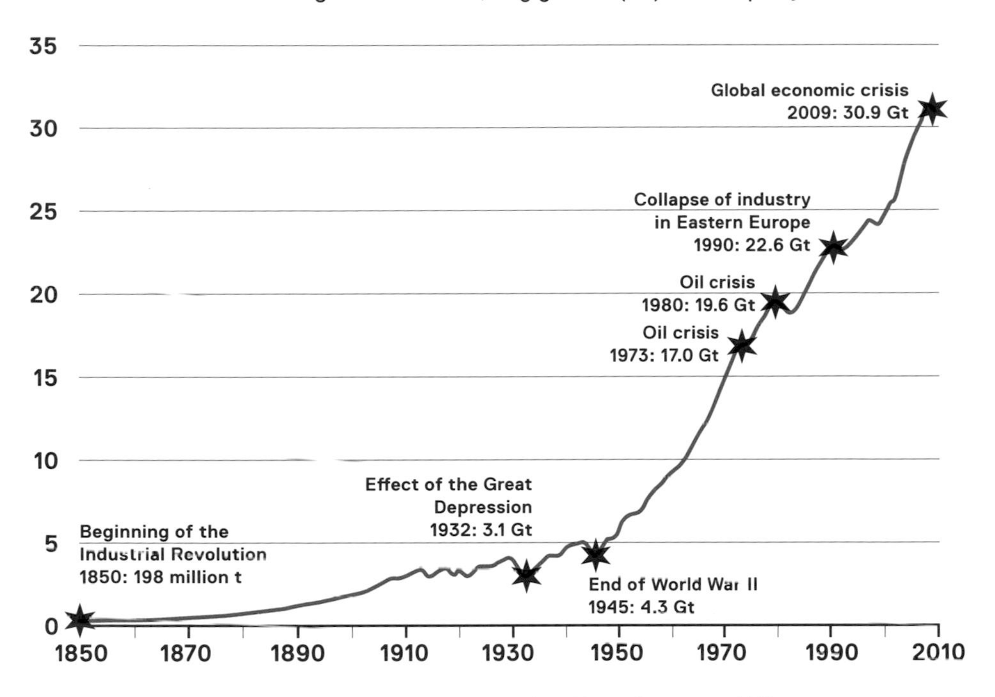

Physicist Joseph Fourier discovered the greenhouse effect as far back as 1824, and chemist and physicist Svante Arrhenius described the effect of carbon dioxide on global warming for the first time in 1896. Since then, these mechanisms have long been experimentally and empirically proved.

It might seem surprising at first that greenhouse gases should have such a major impact on Earth's thermal balance, considering that they are present in such negligible quantities. For instance, the proportion of carbon dioxide in the atmosphere in 2009 was only 0.0386 percent, that of methane 0.00018 percent, and that of nitrous oxide a mere 0.0000322 percent. Or in climatology language: the concentration of carbon dioxide was 386 ppm (parts per million), of methane 1800 ppb (parts per billion), and nitrous oxide 322 ppb.

Direct data on CO_2 levels in the lower layers of the atmosphere have only been collected since 1958 when climatologist Charles Keeling set up a measuring station on Mauna Loa in Hawaii. The Keeling Curve →p. 147, indicating average levels of CO_2 concentrations, continues to the present day and has been corroborated by several dozen other measuring stations around the world. Its baseline has steadily gone up over decades and is superimposed by seasonal fluctuations, with levels decreasing in spring when vegetation needs CO_2 for growth. The Keeling Curve proves that the concentration of CO_2 did indeed increase from 315 ppm in 1958 to 386 ppm in 2009. It also shows that the rate at which concentrations are rising is accelerating. Until the end of the 1980s, the average annual increase was around 1.2 ppm, in the 1990s it went up to 1.5 ppm, and since 2000 it has actually been about 2 ppm per year.

To put this into perspective, CO_2 levels in the atmosphere were only 280 ppm before the age of industrialization. This is documented by the analysis of tiny air bubbles in ice cores. These ice cores also show that CO_2 levels have never been as high as they are today, at least not in the last 850,000 years.

But how do we know this increase is an outcome of anthropogenic CO_2 emissions? This question is not easy to answer since the carbon dioxide cycle includes a number of different carbon and carbon dioxide exchanges between the biosphere, the oceans, and the atmosphere, all of which influence CO_2 levels in the atmosphere in different ways.

Carbon is one of our planet's most important chemical elements. It can be found everywhere on Earth—in its core, its oceans, in the atmosphere, and in plants, animals, and human beings. All living organisms are made up of carbon compounds. Its chemical nature allows carbon to build very complex molecules; of all known chemical elements, it displays the widest variety of chemical compounds. One of the simplest compounds is carbon dioxide, a gas made up of one carbon atom and two oxygen atoms—consequently, one gigaton of carbon is equivalent to roughly 3.67 gigatons of carbon dioxide. Moving through different spheres, carbon atoms can detach from one chemical compound and bond with other elements to form new compounds.

Earth's exact carbon content is unknown. The amount of carbon in the lithosphere is estimated at around 75 million gigatons. Most of it is stored in marine sediment and sedimentary rock. A much smaller share is in the soil, the oceans, and, at least for a short time, in vegetation. Only a fraction of the total carbon in all these spheres, just under

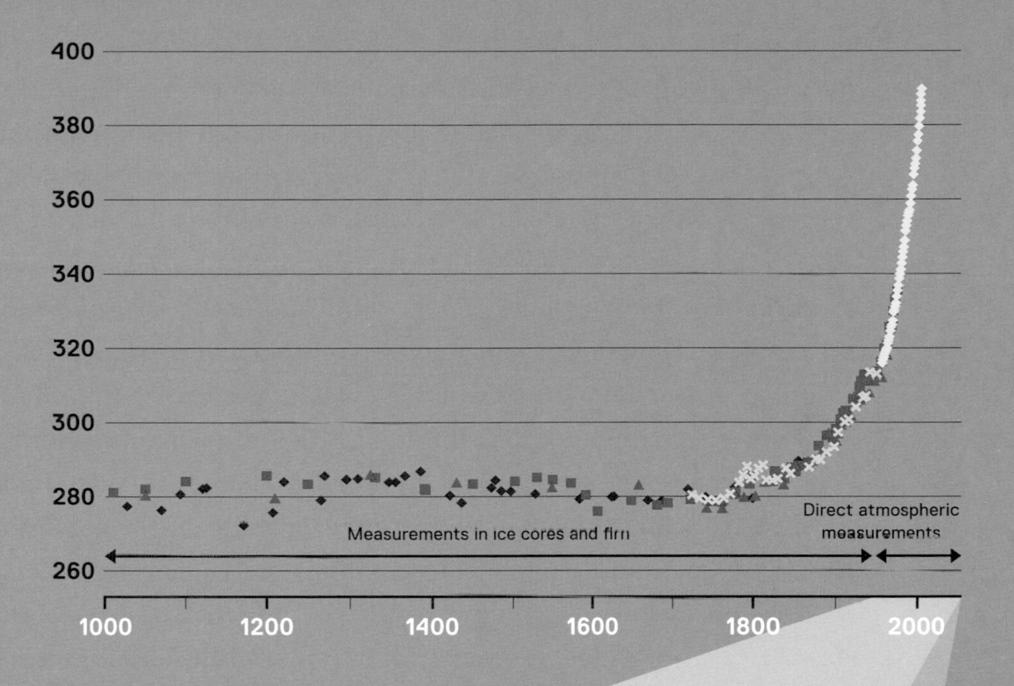

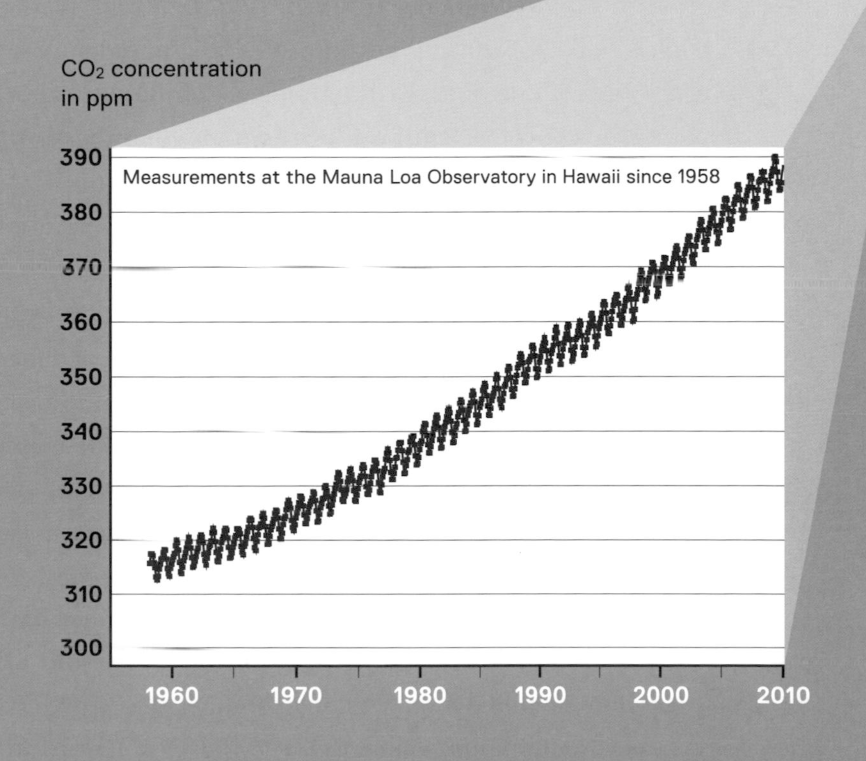

Source: J. L. Sarmiento and N. Gruber, Physics Today 55, no. 8, 2002 (updated 2011)

0.0003 percent or 200 gigatons, is exchanged within and between the different spheres in a year's time →p. 150. It is absorbed by plants through photosynthesis, by animals and humans through the food chain, bound in carbon compounds and released again as carbon dioxide, then absorbed by the oceans and again released, and so on and so forth. The amount of carbon coming from anthropogenic CO_2 emissions is currently around 9.4 gigatons per year, which is only about 5 percent of the carbon in annual circulation.

We know from climate history that CO_2 levels in the atmosphere have varied markedly since the formation of Earth. Rock weathering extracts CO_2 from the air and carries the carbon it contains down into ocean sediments. The other way around, carbon is released back to the atmosphere through volcanic eruptions. This geological cycle contributed over billions of years to the gradual decrease in the atmosphere's carbon dioxide content from roughly 10 percent to today's level of far below one part per thousandth. But this geological cycle is hardly relevant with regard to short-term circulation, periods of 100 or 200 years, since it doesn't process more than 0.1 gigatons of carbon per year.

The cycle driven by plant photosynthesis is much more significant. Terrestrial plants consume about 120 gigatons of carbon per year, but roughly half is released back into the atmosphere by plant respiration within just a few weeks. The other half ends up in the soil in falling leaves and other plant debris, where it is absorbed by microbes and fungi and released again within a year—this process is called soil respiration. Fires have the same effect since the burning of organic materials also releases the carbon they contain.

In the end, only a tiny fraction of carbon remains stored in organic material over longer periods, in wetlands, for example, and in plant deposits that turn into coal, gas, and oil over millions of years. If the terrestrial and marine biospheres always worked unvaryingly in this way, small amounts of carbon dioxide from the atmosphere would gradually be absorbed and stored for centuries or millennia. In fact, the biological carbon cycle did significantly contribute to reducing CO_2 levels in the atmosphere over hundreds of millions of years.

But this has essentially changed since the advent of industrialization. Deforestation and other land-use changes alone have substantially diminished vegetation cover around the globe, releasing about 150 additional gigatons of carbon (or 550 gigatons of carbon dioxide) into the atmosphere. Although this amount may seem small compared to the total volume of roughly 2,300 gigatons of carbon stored in the biosphere, it is in fact sizable compared to the roughly 821 gigatons present in the atmosphere in the form of carbon dioxide. Had these 150 gigatons remained in the atmosphere until now, CO_2 levels would have risen by about 70 ppm. (This calculation does not include the more than 400 gigatons that have also been released into the atmosphere from the combustion of fossil fuels since industrialization began!)

That this 70 ppm increase didn't happen is due to another part of the carbon cycle—the exchange of CO_2 between the atmosphere and the oceans. This exchange works in both directions, yet currently the oceans, along with forests, are the most important natural CO_2 sinks. When atmospheric CO_2 levels rise, oceans absorb some of this carbon dioxide and transport it to the deep sea where some of it remains for hundreds of years. Oceanographers call this effect the physical CO_2 pump.

The oceans store altogether sixty times as much carbon as the atmosphere does, mostly in the form of dissolved CO_2, carbonic acid, bicarbonate, and carbonate. The absorption capacity of the oceans is enormous—over the course of several thousand years they will eventually take up about 80 percent of anthropogenic CO_2 emissions. But the slow rate of ocean transport limits current oceanic uptake to about one-fourth of the annual anthropogenic emissions, i.e. about 2.2 gigatons (net) of carbon per year.

The amount of carbon that oceans can absorb also depends on the climate. As oceans warm up, two things happen: less CO_2 is absorbed from the atmosphere and ocean currents slow down, with the outcome that less CO_2 is transported to deep waters. Finally, the increased input of carbon compounds results in ocean acidification, which in turn diminishes the ocean's absorption capacity.

The same is true of the terrestrial biosphere. The biosphere may be able to absorb less CO_2 when ecosystems change in response to climate change; i.e., when deserts, steppes, and tundra spread, or when deforestation and reforestation, agricultural expansion, and

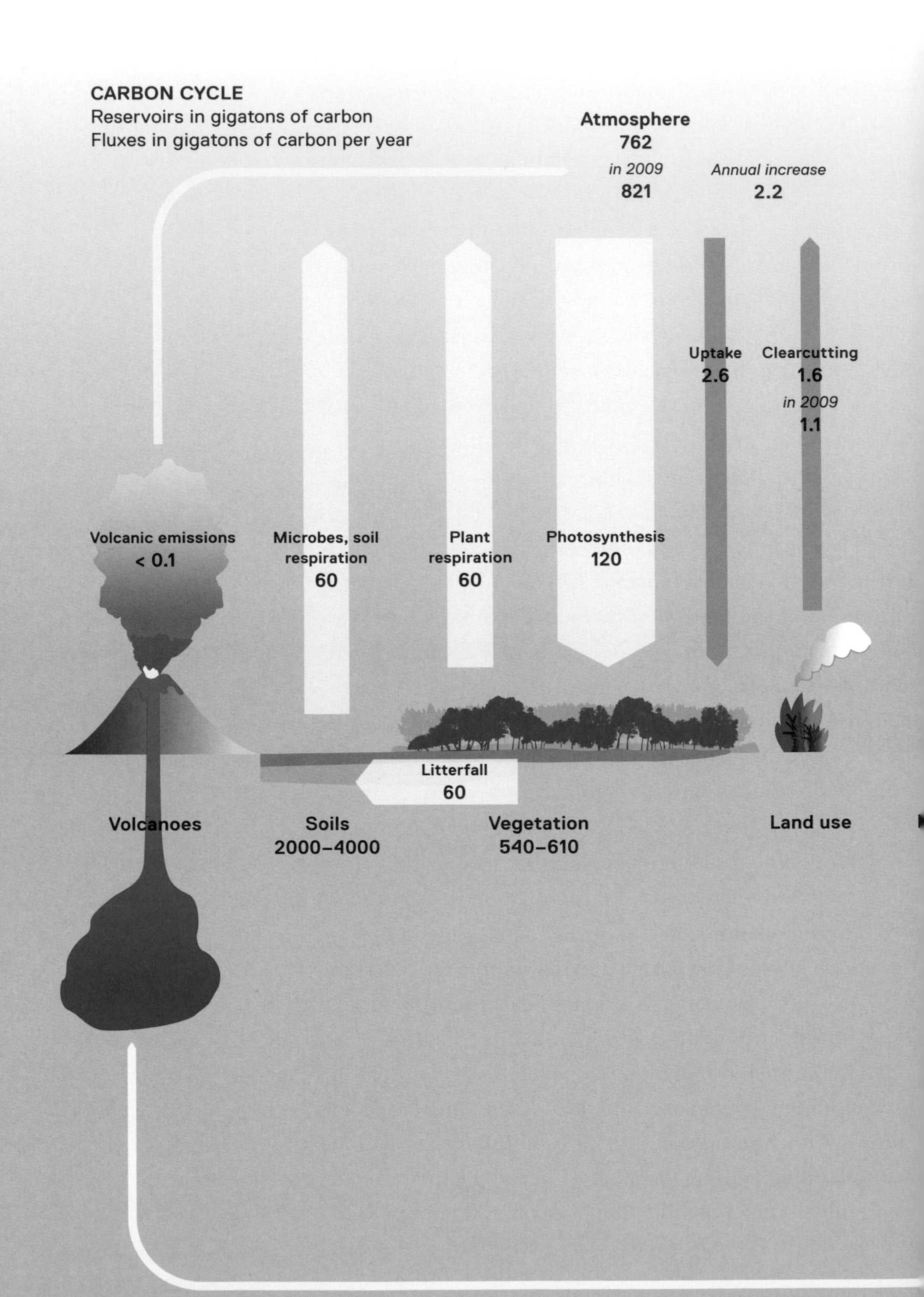

Figures are averages for the decade 1990–2000
Source: 2007 IPCC Report

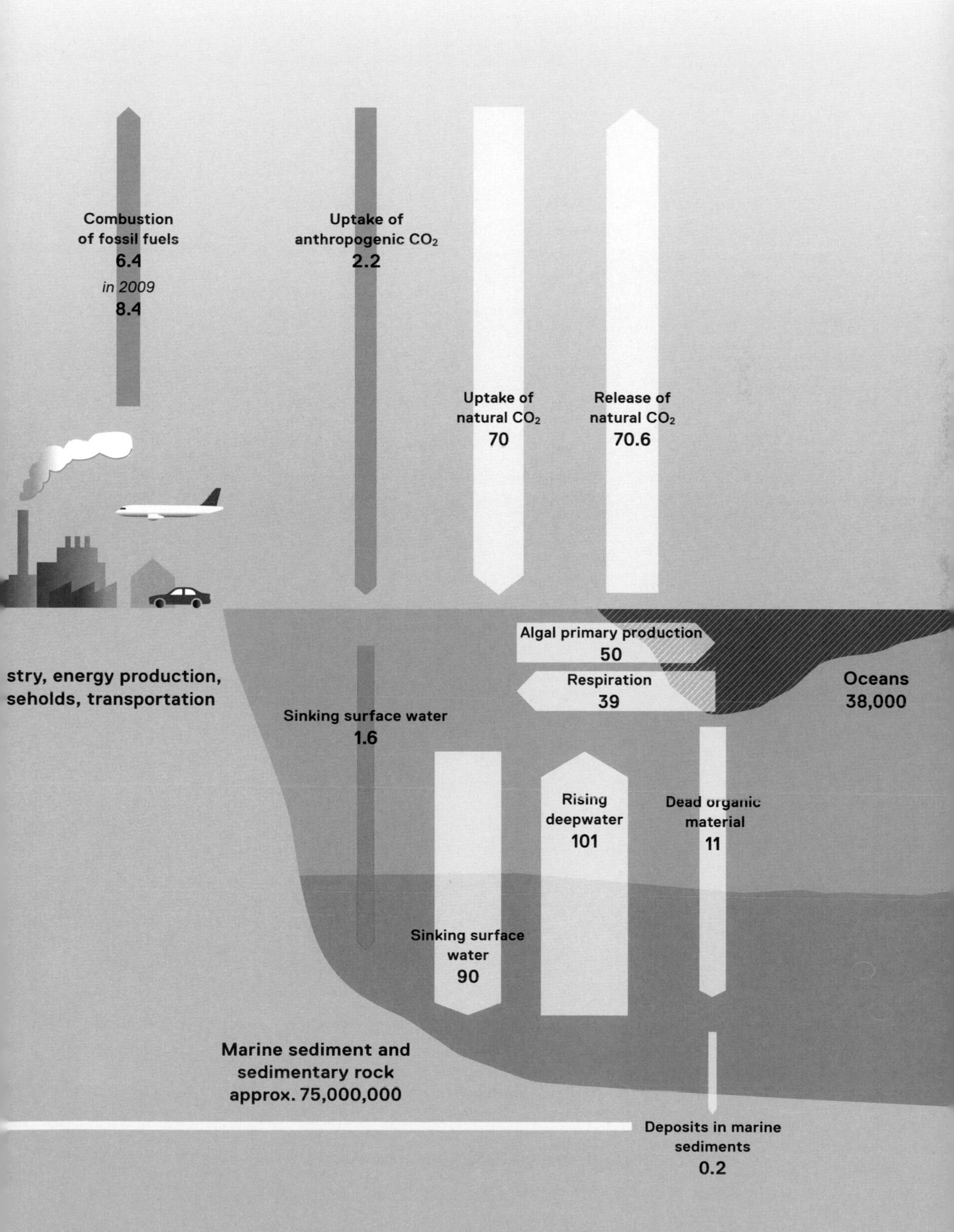
Combustion
of fossil fuels
6.4
in 2009
8.4
Uptake of
anthropogenic CO_2
2.2
Uptake of
natural CO_2
70
Release of
natural CO_2
70.6
Algal primary production
50
Respiration
39
stry, energy production,
seholds, transportation
Oceans
38,000
Sinking surface water
1.6
Rising
deepwater
101
Dead organic
material
11
Sinking surface
water
90
Marine sediment and
sedimentary rock
approx. 75,000,000
Deposits in marine
sediments
0.2

other human activities alter landscapes. Many climate scientists agree that the ability of the oceans and the terrestrial biosphere to absorb carbon dioxide will decrease with global warming. Some even think it likely that the terrestrial biosphere could itself become a source of CO_2 within just a few decades.

But oceans and forests are still effective CO_2 sinks, meaning they are not accountable for the unusually rapid increase of CO_2 levels in the atmosphere. The only source of CO_2 that has clearly increased in the past 250 years is the CO_2 from anthropogenic emissions.

Radiocarbon dating is one method of finding evidence for the combustion of fossil fuels being almost exclusively to blame for the increase in CO_2 emissions. It allows researchers to draw conclusions about the origin of CO_2 molecules in the atmosphere. The process relies on the differing number of neutrons found in carbon atoms. While "normal" carbon (carbon-12) is stable, the less frequently occurring carbon-14 is radioactive and has a half-life of 5,730 years. Since coal, oil, and gas are several million years old, the carbon dioxide they emit during combustion no longer contains any carbon-14. Climate physicists make use of this to calculate the share of carbon from fossil fuels in the atmosphere. Thanks to this process, nuclear physicist Hans Suess was able to prove as long ago as the 1950s that nearly all of the increase in atmospheric CO_2 levels in the last 250 years is anthropogenic in nature.

As described earlier, various exchange fluxes in the carbon cycle ensure that a large share of anthropogenic CO_2 emissions released into the atmosphere every year is removed again sooner or later. On the long-term average, only about 44 percent of CO_2 emissions presently remain in the atmosphere; about 30 percent are absorbed by the oceans, and about 26 percent by the terrestrial biosphere. More precisely, roughly half of every ton of CO_2 emitted is removed within thirty years and another 30 percent in the course of several hundred years, still leaving 20 percent in the atmosphere for many thousands of years. Of the 34.8 gigatons of CO_2 (or 9.4 gigatons of carbon) currently emitted every year, some 17 gigatons of CO_2 (or 4.6 gigatons of carbon) remain in the atmosphere for at least one hundred years and 7 gigatons persist for several thousands of years. Even if we reduce yearly emissions now, CO_2 levels will stay high for many, many years to come.

But carbon dioxide is not the only greenhouse gas whose levels have increased since the advent of industrialization. Atmospheric methane has more than doubled from an estimated 400 to 700 ppb to today's level of 1,800 ppb. The main sources of anthropogenic methane are cattle, rice paddies, waste disposal sites, and large dams. Methane is also released from wetlands, tropical rainforests, mangrove forests, and melting permafrost. Researchers have yet to determine precisely how large the anthropogenic share of the total increase in methane is, although current estimates lie at about 75 percent. Methane gas is retained in the atmosphere for about ten years.

The level of nitrous oxide has gone up the least since industrialization began. The concentration recorded in 2010 was 322 ppb, about 18 percent higher than the preindustrial level. Nitrous oxide is released mostly from the use of fertilizers and when biomass is burned. Researchers estimate that roughly 40 percent of nitrous oxide emissions are anthropogenic.

Finally, greenhouse gases also include a range of trace gases such as fluorinated and chlorinated hydrocarbons (CFCs), most of them man-made. The emission of these compounds rose very rapidly during the 1970s and 1980s, and in response to the danger they pose to the ozone layer, parties to the Montreal Protocol, an international environmental treaty signed in 1987, agreed to halve the production of the most important CFCs by 2000. Follow-up conferences have been successful in achieving an extensive reduction in the production of these gases since 1995 →p.452, and overall levels in the atmosphere have receded slightly since then.

These trace gases have a sizable impact on the greenhouse effect even though their concentrations are far lower than those of carbon dioxide. Per unit, they are more potent than carbon dioxide—methane by 25 times, nitrous oxide by 298 times, and some CFCs by up to 14,800 times. To enable a comparison of greenhouse gases, their impact is usually converted into CO_2 equivalents, found by multiplying their concentration levels by an effectiveness factor. Taking into account the actual concentrations and these effectiveness factors, researchers estimate that methane contributes about 20 percent, nitrous oxide about 5 to 6 percent, and CFCs about 10 percent to the anthropogenic greenhouse effect.

But how do all these carbon and carbon dioxide exchanges between the biosphere, the oceans, and the atmosphere influence global warming? Climate scientists have developed a two-step method to find an answer to this question. Radiative forcing →p.155 shows how much the Earth's radiation balance changes as a result of rises or falls in various factors. Climate sensitivity is meant to provide information on how changes in the radiation balance will affect expected temperature increases.

Radiative forcing is a measurement of the imbalance in the radiative budget caused by changes in greenhouse gases and aerosols. According to the 2007 IPCC report, net radiative forcing increased to a total of 1.6 watts per square meter between the onset of industrialization and 2005. The biggest increases are caused by the long-lasting greenhouse gases: carbon dioxide at 1.66 watts per square meter, and methane, nitrous oxide, and halogenated hydrocarbons together at 0.98 watts per square meter. On the other hand, a number of factors have reduced radiative forcing, most importantly the direct effect of aerosols, changes in land use, and cloud albedo—factors that are directly or indirectly due to anthropogenic interference. But several of these factors have not been satisfactorily researched, and there is a relatively large range of uncertainty of 0.6 to 2.4 watts per square meter.

Long before the concept of radiative forcing was developed, many researchers attempted to calculate how increases in CO_2 levels would affect global warming. They concluded, under laboratory conditions, that a doubling of CO_2 concentrations would induce the global temperature to rise by about 1.2 degrees Celsius (2.2°F).

In reality, the situation is much more complicated. The increase in CO_2 levels is only one of several factors influencing the radiation balance, and thereby global warming. Climate scientists developed numerous methods to include all these factors. Some scientists took experimental laboratory data, added the potential range of uncertainty of the remaining factors, and came up with a climate sensitivity value between 2 to 4.5 degrees Celsius (3.6–8.1°F). Other researchers compared similar climate situations in the past that differed only in their level of greenhouse gas concentrations, and arrived at a climate sensitivity value between 3 to 4 degrees Celsius (5.4–7.2°F).

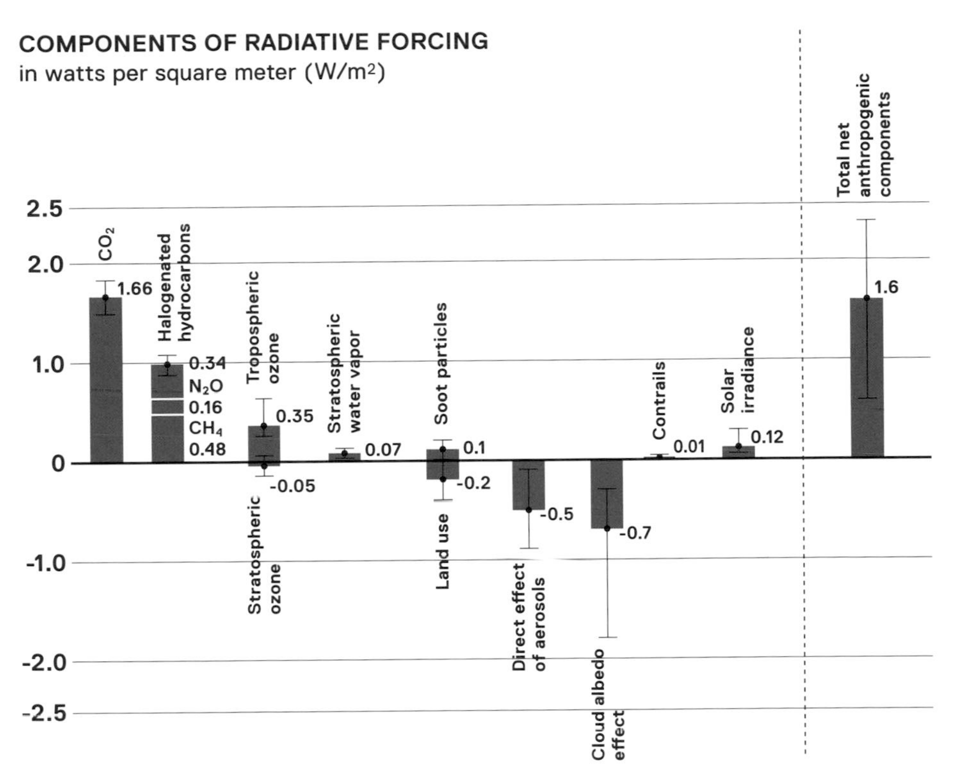

Source: 2007 IPCC Report

More recent studies work primarily with complex computer simulations, testing and varying all of the important influencing factors within their potential uncertainty ranges to find out which of these theoretically computed combinations most precisely reproduces climate evolution during the last glacial period. Since there is a lot of data available on this period, any combinations that don't realistically simulate this ice age can be rejected. Based on information gained from these new models, climate scientists conclude that an increase in radiative forcing of 1 watt per square meter induces a global increase in temperature of 0.8 degrees Celsius (1.4°F). At the current level of radiative forcing of altogether 1.6 watts per square meter, we can expect a rise in the global temperature of 1.3 degrees (2.3°F). The ocean delay effect means that over time another 0.6 degrees (1.1°F) would be added to the 0.7 (1.3°F) degrees observed so far, until the global surface temperature settles into a new equilibrium.

For historical but also largely practical reasons, climate scientists continue to prefer the conventional expression of the relationship between radiative forcing and CO_2 levels. They calculate the value for radiative forcing when preindustrial CO_2 levels are doubled (= 3.7 watts per square meter) and arrive at a prediction of 3 degrees Celsius (5.4°F) for global warming. The formula most commonly used in climatology doubles the CO_2 level from 280 ppm to 560 ppm, the outcome of which is a rise in the average global surface temperature of 1.5 to 4.5 degrees Celsius (2.7–8.1°F), whereby the most probable value is 3 degrees (5.4°F).

Yet new model simulations lead researchers to suspect that climate sensitivity is probably higher. That would mean that a doubling of CO_2 levels could cause a temperature increase of considerably more than 3 degrees Celsius.

This would apply in equal measure to the two-degree Celsius (3.6°F) target. Climate scientists have calculated that there is a 50 percent probability that CO_2 levels at 450 ppm would lead to a two-degree rise in temperature. But if the suspicion is confirmed that climate sensitivity is actually higher, or if we want to be more than 50 percent certain that the increase in global temperature will not exceed 2 degrees, then these estimates dictate that CO_2 levels should not be allowed to rise above 400 ppm for longer periods of time. However, CO_2 levels had already reached 388 ppm by the end of 2010. This value increases every year by about 2 ppm. No matter how we look at it, there isn't much time left for us to meet the two-degree target set at the climate conference in Cancún.

THE FUTURE OF EARTH'S CLIMATE IS IN OUR HANDS

The advent of industrialization marked the point at which the human race itself became an important climate factor. The invention of the steam engine and the internal combustion engine, the use of machines to mine and industrially process natural resources, the mass production of goods, the industrialization of agriculture, the rapid rate of global population growth, rural depopulation, urbanization, and globalized transportation brought wealth and prosperity to some parts of the world, but also led to immense interference in the natural environment. This includes first and foremost our enormous consumption of fossil fuels, but also deforestation and the large-scale disruption of ecosystems and changes in vegetation caused by irrigation, melioration, and changes in agricultural land use, and the pollution of air and water with aerosols and chemicals. Consequently, since the beginning of global industrialization, atmospheric carbon dioxide levels have risen ten times more rapidly than in the past 22,000 years since the last glacial period. The composition of air during those millennia is well researched, based on information gathered from ice cores. There is hardly any reasonable doubt that humankind has had a decisive impact on the development of Earth's climate in the past 250 years.

Since the beginning of industrialization, the average global temperature has increased by 0.7 degrees (1.3°F) due to anthropogenic greenhouse gas emissions. Even if carbon dioxide concentrations were to level off at 386 ppm, the temperature would continue to rise by 0.5 to 0.6 degrees (0.9–1.1°F) over coming decades until Earth reached a new balance. If we want to meet the two-degree target agreed upon at the Cancún climate conference in December 2010, CO_2 levels should not exceed 400 ppm for prolonged periods of time. Should concentrations go up to 560 ppm, which is twice the preindustrial level, the average global surface temperature will increase by 3 degrees (5.4°F), probably considerably more.

Climate scientists cannot provide us with very reliable projections on how climate will develop during this century. Even if they knew and could calculate every single detail in all climate processes, this would be an impossible task, the reason being that the most important climate factor is also the least predictable—humankind. No climate scientist or economist or politician can tell us how much fossil fuel will be burned in the future, whether or not the destruction of rainforests will continue on the same scale, how environmental technologies will develop, what the future of the global economy will be, and whether and to what extent the global community will ultimately commit itself to effective climate protection measures. All this depends on developments that are impossible to predict and on decisions that have yet to be made. Only two or three decades ago, no one could have predicted that structural reforms in the Chinese economy would turn China into the world's biggest greenhouse gas emitter within just twenty years.

The IPCC's first climate assessment in 1990 still very simplistically assumed that fundamental socioeconomic development would

Anthropogenic CO_2 emissions have increased fourfold over the last fifty years. The ten warmest years of this period have all been recorded in the last fifteen years.

continue more or less unchanged. But it was clear even then, shortly after the unexpected collapse of socialist regimes in Eastern Europe, that a business-as-usual approach described only one of many potential eventualities.

Scientists therefore chose a much more diversified approach for the *Special Report on Emissions Scenarios* (SRES), which was also used for the *IPCC Fourth Assessment Report* of 2007. They designed a number of different scenarios to predict how the world might develop over the next few decades. Each of these scenarios is based on different assumptions about a variety of important socioeconomic factors: future global population growth, technological advances, whether society puts more emphasis on economic development or on environmental sustainability, whether the world continues to be divided into completely heterogeneous regions with varying degrees of development, or if globalization results in fewer regional differences in the distribution of wealth, and, finally, how humans manage fossil fuels and renewable energy sources.

Experts analyzed a total of thirty-five SRES scenarios and chose six scenario groups that represented four fundamentally different development paths. Their choices had nothing to do with the likelihood of the scenarios or how climate protection measures currently under debate would affect climate change; rather, the scenarios were meant to show how various widely diverging development trends would affect the climate.

The many factors that were analyzed can be grouped largely into two fundamentally different categories: the A scenarios, which describe a world in which economic growth continues to be a driving force; and the B scenarios, which describe a world that focuses on the need for environmental sustainability. These two categories are subdivided into two more groups: 1 stands for the future development of a more homogenous world with diminishing differences in the distribution of wealth; and 2 stands for a heterogeneous world characterized by considerable regional differences.

A1 scenarios describe a globalized world with a steady level of strong economic growth, rapid advances in new and efficient technologies, and a growing global population that peaks at nine billion by 2050. These scenarios assume that economic and cultural differences between the world's regions will decrease, and that income and standards of living will align with each other at a relatively high level. This A1 group was divided into three scenarios which took energy sources into consideration: A1FI (fossil fuel intensive) describes a world that continues to depend on the intensive use of fossil fuels, A1T (predominantly non-fossil fuel) describes a world that rapidly shifts to the use of renewable energy sources, and A1B (balanced) that stands for a mixed use of fossil fuels and renewable energy sources.

In contrast, A2 scenarios describe a heterogeneous world focusing on rapid economic growth, but in which considerable differences between industrial nations and emerging and developing countries continue to exist. The IPCC somewhat euphemistically describes this in its report as "self-reliance and preservation of local identities," but surely it would not be entirely wrong to characterize this heterogeneity as a continuation of great social injustice and great economic disparity. In these scenarios, the global population continues to grow past the middle of the century, possibly rising to fifteen billion; economic growth, per capita income, and technological advances develop differently from region to region and vary greatly.

B1 scenarios represent a globalized world with diminishing regional differences and a gradual decline in global population after 2050, as in A1. The difference to the three A1 scenarios is that in this scenario, economies and society seek ecologically sustainable solutions; this world develops toward a service and information society with declining material intensity and increasingly clean and efficient technologies.

Finally, B2 scenarios describe a heterogeneous world in which developments occur regionally, but with a focus on economic, social, and environmental sustainability, unlike the A2 scenarios. The global economy in these scenarios is characterized by moderate growth, and technological change takes place on a regional level and is adapted to local needs. The global population continues to grow, rising to roughly ten billion by 2100.

Although the B scenarios do not explicitly include the political climate-related intervention called for in today's debates, many of their underlying assumptions imply the implementation of climate protection measures. Putting "emphasis on environmental sustainability" is simply another way of saying that these measures should be installed.

Using these socioeconomic scenarios, climate scientists have calculated how climate may develop in this century →p.162, and have reached the following conclusions:

All scenarios describe a continued increase in greenhouse gases until about 2030. This increase, measured in CO_2 equivalents, will range between 25 and 90 percent above 2000 levels, depending on the scenario. The consumption of fossil fuels will rise by 40 to 110 percent, with CO_2 emissions rising correspondingly, while emissions of other greenhouse gases will level off or even decline in some scenarios.

But differences between the scenarios become more pronounced after 2030. Climate scientists calculate that by the end of the century, the most environmentally friendly scenarios B1 and A1T will experience a decline in emissions of up to 25 percent below levels in 2000. In contrast, models predict that scenarios A1F1 and A2 will see a continuous and steep increase of emissions of about 200 percent.

Obviously, such socioeconomic projections include a wide range of uncertainty. The invention and development of new technologies, future demand for consumer goods, and the rise in the private use of motorized transportation in emerging nations, to name just a few factors, are very difficult to predict. Moreover, climate sensitivity, the base for calculating temperature increases, is still relatively imprecise with a range of uncertainty between 2 to 4.5 degrees Celsius (3.6–8.1°F). For this reason, the scopes of various SRES scenarios overlap. When temperature anomalies are calculated relative to preindustrial times, all SRES scenarios exceed the critical two-degree warming target, which experts have declared to be the maximum acceptable amount of global warming. However, in addition to the SRES scenarios described in the IPCC report, there are several economic models that take into account potential political measures aimed at reducing greenhouse gas emissions; these come very close to meeting the two-degree target. With a view to the next IPCC assessment report, a large, international group of climate scientists has designed new scenarios intended to enable more precise statements

SRES SCENARIOS FROM 1900 TO 2100

Surface temperatures in degrees Celsius* relative to 1980 to 1999

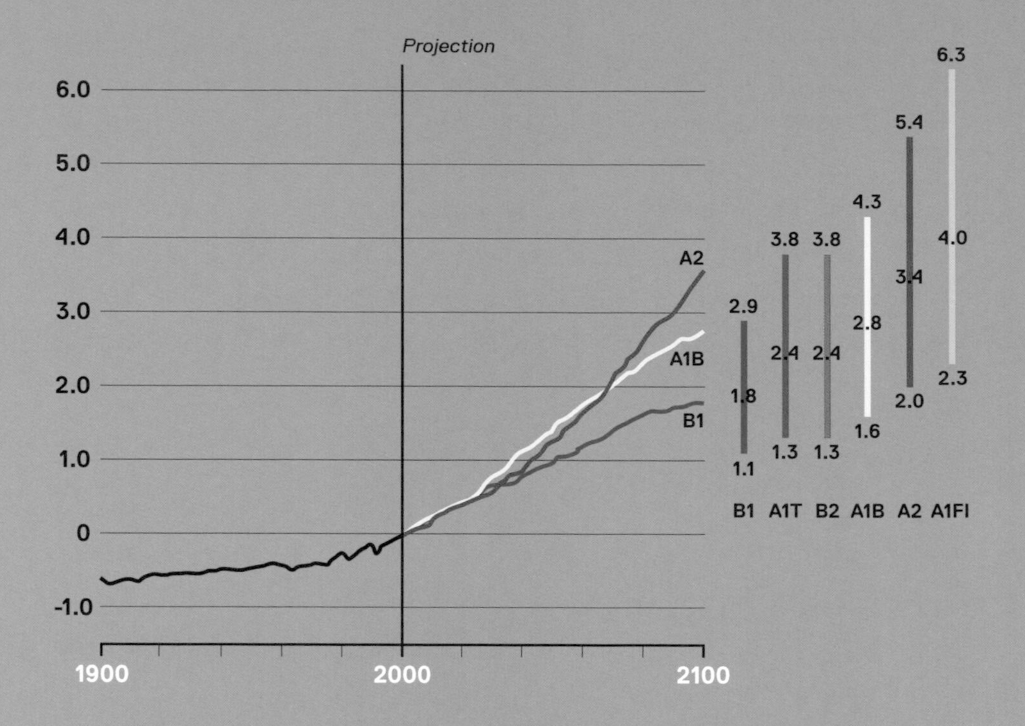

A1 scenario family
A globalized world with strong economic growth and rapid advances in new, efficient technologies. Global population grows to 9 billion by 2050, with income and living standards converging at a high level. A1FI: intensive use of fossil fuels; A1T: predominantly renewable energy consumption; A1B: mixed use.

A2 scenario family
A heterogeneous world with rapid population growth, but differences in economic growth and slower technological advancement.

B1 scenario family
A globalized world with population growth as in A1, but economies and society seek ecological and sustainable solutions.

B2 scenario family
A heterogeneous world with moderate population and economic growth; solutions are found at the regional level and adapted to local needs.

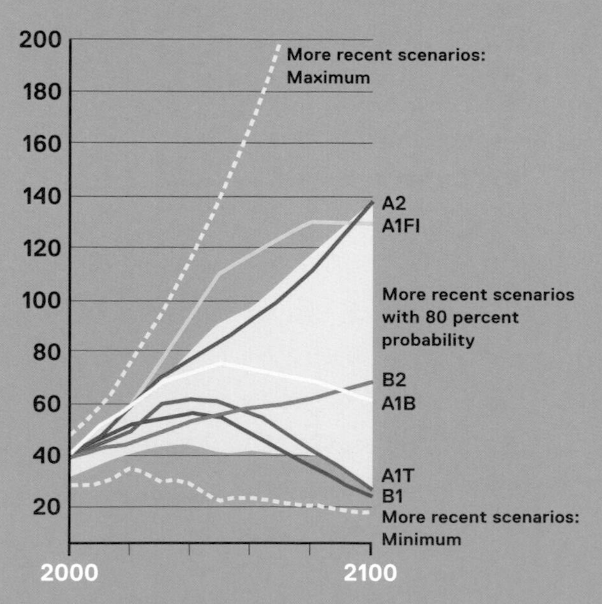

Source: 2007 IPCC Report
* Note: 1 degree Celsius equals 1.8 degrees Fahrenheit

on the connection between socioeconomic development and climate change. The Representative Concentration Pathways (RCP) project is meant to provide information of the kind needed for decision-making on climate policy for the near term, from now until 2035, and for the long term up to 2100, and even up to 2300.

But even the current SRES scenarios allow us to reach some clear conclusions. None of the scenarios meets the two-degree target without the help of additional climate policy measures, not even if the use of fossil fuels is largely abandoned, as in scenario A1T. Added to that, the most recent measurement data indicate that developments in temperature are more likely to take place at the upper end of estimated ranges. Finally, the SRES scenarios show that a globalized world geared toward rapid economic growth with continued reliance on fossil fuels is likely to experience global warming of 4 degrees Celsius (7.2°F) or more.

Climate scientists have looked not only at the development of global temperatures, but also explored another consequence of climate change—for example, the rise in sea levels. Depending on the IPCC scenario referred to, sea levels are expected to rise by 18 to 59 centimeters by the end of the century. This forecast is based, however, on the assumption that polar ice masses will continue to melt at the present rate. But new studies show that the climate in the northern polar region is warming at a much faster rate than the global average. Many climate scientists suspect that by the end of the century sea levels will have risen much more than predicted in the SRES scenarios.

Climate science has made huge progress over the past few decades. Today this science can explain all of the essential mechanisms regulating the global climate system and calculate them with an acceptable degree of reliability, even though some processes are not yet fully understood, and various computer models and simulations sometimes provide us with diverging results. The IPCC reports, which assess the latest scientific information related to climate research, take this into account by wording their conclusions very carefully and specifying the margins of uncertainty. But even the best climate models cannot help climate scientists to make reliable predictions about the most unpredictable factor—future human activity.

Scientists can describe how human activities, our behavior, will affect the climate. They can also tell us what we need to do and what measures we can take to influence climate change. But in the end, the conclusions we draw from this information and what course we set for the climate of the future is left for us to decide and act on together.

Digging for gold in the Amazon Basin — agricultural expansion is not the only reason rainforests are cut down. Novo Aripuanã, Brazil, 2007. *Paulo Santos/Reuters*

Commercial interests are the biggest threat to tropical rainforests; international climate agreements could end deforestation. Lagos, Nigeria. *Stuart Franklin/Magnum Photos*

More harm than good — rainforests are being replaced by oil palm plantations for biofuel. Ketapang District, West Kalimantan, Indonesia, 2010. *Crack Palinggi/Reuters*

Harvesting soy in Mato Grosso, Brazil—monoculture farming reduces biodiversity and impedes the free exchange of water and gases between soil and atmosphere. *Paulo Whitaker/Reuters*

Rice cultivation accounts for about 17 percent of global methane gas emissions. Sichuan Province, China. *Pierre Colombel/Corbis*

Lots of cows, lots of methane — livestock breeding accounts for about 39 percent of global methane emissions. Cattle ranch in Holdrege, Nebraska, USA. *Nati Harnik/Keystone*

53
53

Smog and a massive cloud of dust from the Sahara cover the Greek capital of Athens, with the Acropolis disappearing in the haze. *Simela Pantartzi/Keystone*

A cold winter's morning in Zurich, Switzerland. *Andreas Meier/Reuters*

More than 4.7 million cars were registered in Beijing by the end of 2010. *Reinhard Krause/Reuters*

南 S
人
阜成门
平安里
PINGANLI

The uplift factor—exhaust gases emitted by aircraft at cruising altitudes have a marked additional negative impact on the climate. Tokyo, Japan. *Toshiyuki Aizawa/Bloomberg/Getty Images*

Manufacturing aluminum is one of the most energy-intensive industrial processes. Azerbaijan, 1993.
Ria Novosti/Science Photo Library/Keystone

Coking coal is partly responsible for the cement industry's unfavorable climate balance.
Shanghai Coking and Chemical Plant, China. *Stuart Franklin/Magnum Photos*

Iron ore smelting and iron and steel production require high temperatures and generate large amounts of CO_2 emissions. Steelworks in the United States, 2004. *Chris Steele-Perkins/Magnum Photos*

Cement mills account for about 5 percent of all anthropogenic CO_2 emissions; new technologies could cut this figure in half by 2050. Nobeoka City, Japan, 2000. *Mark Power/Magnum Photos*

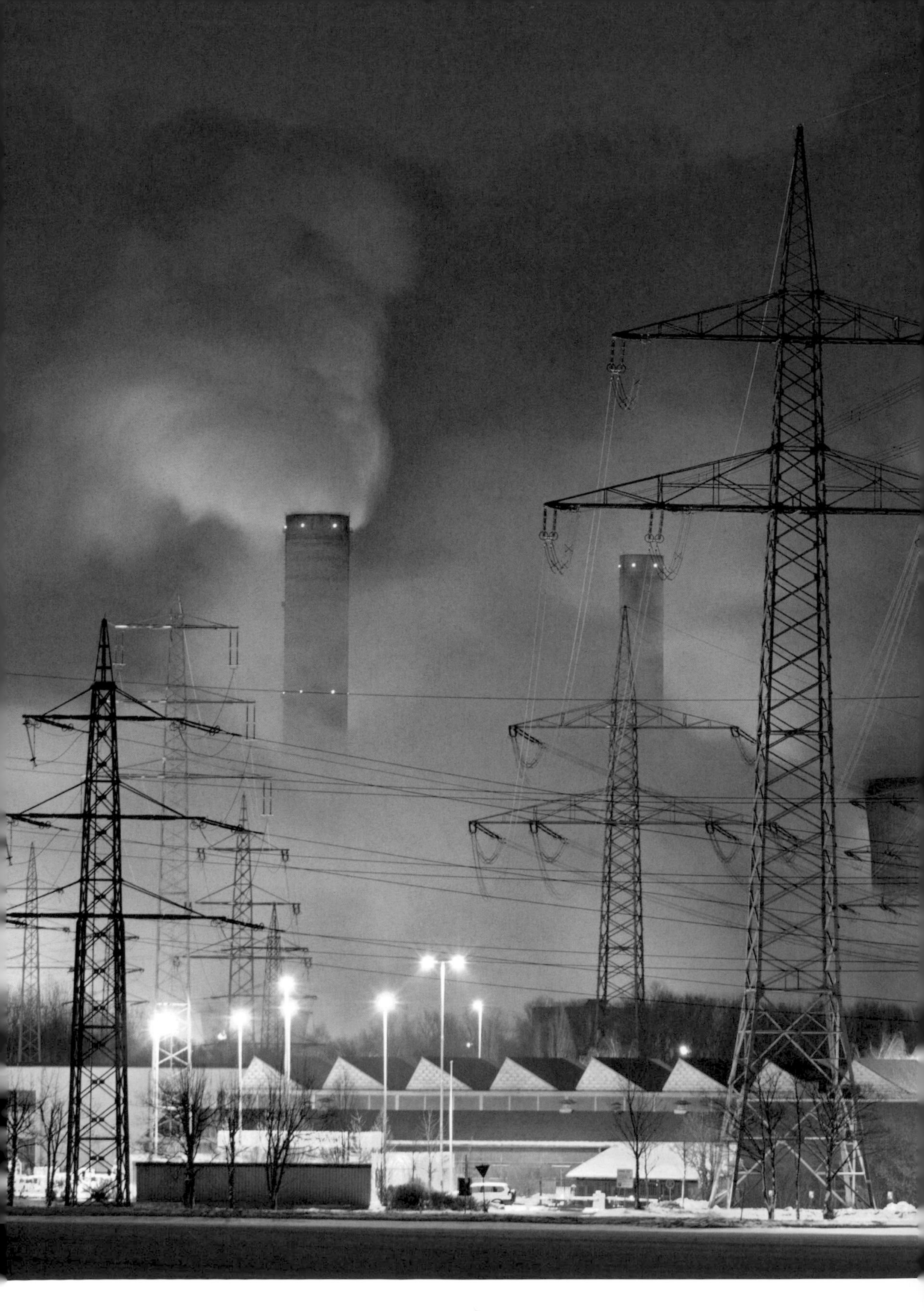

Energy demand beyond limits — lignite-fired power plant in Weisweiler, Germany, 2010. *Paul Langrock/LAIF*

EXTREM
WEATHE
EVENTS
2010

E

R

OF

“Apart from non-weather-related natural disasters such as earthquakes, the frequency of weather-related events such as windstorms, rainstorms, and floods has also increased, claiming the lives of tens of thousands of people and causing enormous damage. Of a total of 950 natural disasters recorded in 2010, nine-tenths were weather-related events.”

From the German Press Agency (dpa) dispatch on the Munich RE Group’s press release reviewing 2010 natural disasters, January 3, 2011.

MONGOLIA, FEBRUARY/MARCH 2010

At the beginning of the year, Mongolia was gripped by a "white zud." That is the Mongolian term for winter weather spells with heavy snowfalls when livestock no longer find fodder. The temperature sank to below minus 40 degrees Celsius (-40°F), and for weeks 80 percent of the land lay under a deep layer of snow. About 4.4 million cattle, sheep, goats, horses, and camels starved or froze to death—10 percent of the livestock in the country. The zud meant a serious economic setback for the country's rural population. Hujirt Sum, Mongolia. *Ganbat Namjilsangarav/Keystone*

SICILY AND CALABRIA, ITALY, FEBRUARY 2010

"I saw my house just slip away," reported a villager from San Fratello in Sicily. Like him, several thousand people in Sicily and Calabria had to flee their homes after entire hillsides collapsed following torrential rainfall. In southern Italy, the danger of landslides is acute because the vegetation that holds the soil in place is often burned off and houses are built on steep, unstable ground. Just a few months earlier, in October 2009, thirty-seven people had already lost their lives near Messina in a mudflow caused by a hillside slide. *Carmelo Imbesi/Keystone*

THE MEKONG RIVER, THAILAND, OCTOBER 2009 TO APRIL 2010

The Mekong River feeds one hundred million people in Thailand, Laos, Cambodia, and Vietnam. It provides water and fertile sediment for the fields on its banks. Mekong fish supply 90 percent of the animal protein in the region. In April 2010, the water level was the lowest it had been in decades, and in some places it was barely forty centimeters deep. Never in the past fifty years had rainfall been scarcer than in the period from October to April. To make matters worse, China began filling the reservoirs of new dams in the Upper Mekong Basin. Fishermen near Nong Khai in northern Thailand. *Dario Pignatelli/Bloomberg/Getty Images*

DRAGUIGNAN, VAR DEPARTMENT, FRANCE, JUNE 2010

The French minister of the interior called it an "unprecedented disaster for the region." The interior region behind the Côte d'Azur saw as much rain within a few hours as it usually does over several months. This triggered flash floods and many people were not able to evacuate to safety in time. There were twenty-five dead, most of them drowned in their vehicles. More than 1,200 people had to be rescued from the disaster scene by helicopter. According to the meteorological service Météo France, this was the worst flooding to have affected these villages in almost two hundred years. *Claude Paris/Keystone*

CHINA, JUNE/JULY 2010

At the end of June, a heat wave gripped sixteen Chinese provinces, and at the beginning of July, temperatures in Beijing reached 40.6 degrees Celsius (105.1°F), the highest temperature registered in sixty years. High humidity made the heat almost unbearable; many people suffered from heatstroke. They sought relief in air-conditioned shopping centers or in swimming pools, as seen here in Suining in Sichuan province. Air conditioning was actually installed in the panda bear enclosure in the Shanghai Wild Animal Park. Power consumption in Beijing soared to unprecedented heights during the heat wave. *Stringer/Reuters*

NEW DELHI, INDIA, MARCH TO JUNE 2010

Northern India experienced several heat waves in the spring and summer of 2010 that sent temperatures up to 47 degrees Celsius (117°F). Public life came to a virtual standstill as people avoided going outdoors. Even in the Himalayan resort of Shimla, highly prized as a summer residence by English colonialists for its cooler climate, a new record high was set in April. The lack of vegetation in cities made the heat problem even worse. Travelers in New Delhi seek to cool off on the stone floor of a railroad station. *Saurabh Das/AP/Keystone*

NEW YORK, USA, JULY 2010

In New York, and (indeed) along the entire east coast of the United States, the summer of 2010 was one of the hottest ever, with temperatures reaching 40 degrees Celsius (104°F). The heat drove people in flocks to swimming pools, and air conditioners ran at full blast. Power consumption broke records. Two months later, a heat wave hit California—in downtown Los Angeles the temperature soared to about 45 degrees Celsius (113°F), the highest ever since record keeping began in 1877. *David Goldman/AP/Keystone*

MOSCOW, RUSSIA, JUNE/JULY 2010

In the summer of 2010, central Russia sweltered under an abnormally long and severe heat wave; many regions recorded the highest temperatures in 130 years. Temperatures in Russia during the first weeks of July were 6 to 9 degrees Celsius (11–16°F) above long-term averages. Heat and a lack of rain for weeks parched the ground and vegetation, sparking wildfires in forests and peat bogs. Drought and extreme temperatures lasted until the middle of August. *Mikhail Metzel/AP/Keystone*

CHINA, APRIL TO AUGUST 2010

During the spring and summer, torrential rains caused dams to burst and triggered landslides and severe flooding in several parts of China. Several hundred people died and millions had to flee the deluge. The Yangtze River swelled, reaching record-high floodwater levels that threatened the Three Gorges Dam, the biggest dam in the world. Up to 100,000 cubic meters per second of muddy water had to be released to lower the water level of the reservoir, already filled to maximum capacity. This caused severe flooding along the Yangtze downstream. *Miao Qiunao/Keystone*

PORTUGAL, JULY/AUGUST 2010

In July and August, a heat wave with temperatures above 40 degrees Celsius (104°F) and no rain led to the outbreak of raging wildfires in Portugal. Thousands of hectares of pine, eucalyptus, and cork oak forests, unique and of ecological importance, were devastated. Although Portugal has to cope with forest fires every year, some 600 fires were raging in July 2010, the largest number in five years. Hamlet of Bouça in the municipality of Manhouce near São Pedro do Sul. *Rafael Marchante/Reuters*

RUSSIA, AUGUST 2010

Weeks of heat and drought sparked forest fires in extended parts of Russia. In August, 9,000 square kilometers of woodland and peat bog were on fire; at least fifty villages burned down completely and even cities were threatened. At the beginning of August, Moscow was covered by a thick blanket of toxic smoke. Authorities ordered residents to remain indoors and keep windows and doors shut. According to official reports, the mortality rate in Moscow went up by nearly 70 percent compared to the previous year. Wildfire threatens Vyksa, 500 kilometers east of Moscow. *Nikolay Sinev/Ria Novosti/Keystone*

PAKISTAN, JULY/AUGUST 2010

"It is the worst flooding of all time in Pakistan," stated UN Special Envoy Jean-Maurice Ripert. Following persistent monsoon rains, the Indus River overflowed and inundated one-fifth of the entire land area. Streets and bridges were swept away and more than 1.7 million dwellings were damaged. Millions were forced to flee, as seen here in the district of Muzaffargarh in Punjab. Abdullah Hussain Haroon, Pakistan's ambassador to the UN, estimated that 15 to 25 percent of crops were destroyed. Damage to the infrastructure has significantly set back the country's economic development. *Adrees Latif/Reuters*

VERACRUZ, MEXICO, SEPTEMBER 2010

Hurricane Karl made landfall on the Mexican coast on September 17, blowing sustained winds of up to 200 kilometers per hour. Karl was the sixth Atlantic hurricane of the year. Fourteen offshore drilling platforms had to be evacuated, suspending oil production. Mexico's only nuclear power plant was shut down as a precaution. In the coastal city of Veracruz alone, 30,000 local residents lost their houses due to landslides and flooding. Six more Atlantic hurricanes followed, adding up to nearly as many as during the record hurricane season of 2005. The seafront in Veracruz. *Jorge Reyes/Keystone*

BURKINA FASO, SAHEL REGION, WEST AFRICA, SUMMER OF 2010

About ten million people in the Sahel region faced famine in the spring and summer of 2010. Food storehouses were empty due to the lack of rain and crop failures in the previous year. Rainfall has drastically decreased in the region since 1970. Scientists attribute this partly to the significant warming of the oceans in the Southern Hemisphere. In the seasonally parched riverbed of the Beli River in Burkina Faso, inhabitants dig for edible tubers of the water lily, a traditional source of nourishment in times of drought. *Georg Gerster/Keystone*

BENIN, WEST AFRICA, OCTOBER 2010

After a long spell of drought, torrential rains fell in Benin in West Africa, followed by the worst inundations since 1963. The United Nations estimates that 680,000 of Benin's 9 million inhabitants lost their homes and 40 percent of the harvest was destroyed. In the port city of Cotonou at the mouth of the Ouemé River, residents of lower-lying districts could reach their flooded homes only by boat. *Loetitia Raymond/CARE*

THE PHILIPPINES, OCTOBER 2010

Typhoon Megi was one of the most powerful tropical cyclones ever recorded. Its central barometric pressure fell to 885 millibars. Accompanied by pounding rain, it raged over the Philippines with wind gusts up to 260 kilometers per hour. Thirteen people died and more than 200,000 were left homeless. A similarly powerful typhoon had claimed over 1,000 lives in 2006. This time, the authorities were able to prevent the worst by giving early warning and evacuating inhabitants. Isabela province, the Philippines. *Dennis M. Sabangan/Keystone*

AMAZON REGION, BRAZIL, JULY TO NOVEMBER 2010

In November 2010, the Amazon River held only half the volume of water usual for this time of year. The water level in Manaus fell to its lowest point since measurements began in 1902. Ship traffic and fishing came to a virtual standstill; millions of dead fish rotted on river shores and sand banks. This event was even more severe than the "once-in-a-century" drought that had afflicted the area in 2005. Climate scientists expect more such dry spells in the future—with incalculable consequences for the Brazilian rainforest. *Raimundo Valentim/Keystone*

HAITI, NOVEMBER 2010

In November, Hurricane Tomas inundated parts of the Caribbean island of Haiti. Tomas was the twelfth and last Atlantic hurricane of the 2010 season. It began developing on October 29, making it the latest Atlantic hurricane ever recorded. Just months earlier, on January 12, 2010, an earthquake had wreaked havoc, destroying about nine-tenths of all the buildings in Haiti and killing approximately a quarter of a million people. This photograph was taken the day after the hurricane during an overflight by the US Navy. *US Navy/Keystone*

COLOMBIA, DECEMBER 2010

"What can we do if twenty-eight out of thirty-two departments must be declared disaster areas?" asked Colombian president Juan Manuel Santos in December 2010. His country experienced the worst floods in its history after the rainy season brought exceptionally heavy rainfall. Vast expanses of Colombia were inundated when dams broke; about 2.2 million people lost their homes. The intense rainfall was attributed to the weather phenomenon known as La Niña, which occurs every few years. Here the village of Caucasia in the Department of Antioquia. *Mauricio Duenas/Keystone*

QUEENSLAND, AUSTRALIA, DECEMBER 2010/JANUARY 2011

Queensland state treasurer Andrew Fraser spoke of a "disaster of biblical proportions." Intense, persistent rainfall suddenly ended a twelve-year drought and submerged an area larger than Germany and France combined. Some 200,000 people were directly affected. Commerce and industry were also badly hit, as seen here in an aerial view of an industrial zone in Brisbane. The effects were felt far beyond Australia's borders because coal mines had to be closed due to the floods, leading to a worldwide shortfall in coking coal, indispensable in cement plants. *Tim Wimborne/Reuters*

LONDON, UNITED KINGDOM, DECEMBER 2010

Unusually heavy snowfall forced Gatwick Airport near London to close down twice in December, the first time early in the month for two days and then again for a few hours shortly before Christmas, disrupting the travel plans of thousands of passengers. It is rare for snow to fall so early in winter in the British Isles. The British transportation system was not prepared for heavy snowfall and traffic broke down—numerous trains were canceled and roads were clogged with major traffic jams. *Ian Bartlett/Camera Press/Keystone*

NEW YORK, UNITED STATES, DECEMBER 2010

"A storm for the history books," said Mayor Michael Bloomberg. The biggest storm ever recorded in New York brought snowfall not seen since 1948. "New York was a city of apocalyptic silence in the morning," noted the *New York Times.* There were practically no cars in the streets; snowdrifts brought bus, train, and air service to a standstill. But New Yorkers did not seem to be very impressed. Mayor Bloomberg said, "The city is going fine. Broadway shows were full last night. There are lots of tourists here enjoying themselves." *Anthony G. Catalano*

ISRAEL, NOVEMBER/DECEMBER 2010

2010 was the seventh consecutive year of drought in Israel, with the month of November being the warmest in sixty years. In the northeast, a small fire spread so rapidly that it turned into several extensive forest fires, becoming the worst natural disaster in the country's history. Fires destroyed some five million trees in an area of thirty-five square kilometers—almost one-third of Israel's forested land. Several settlements, here the artists' village of Ein Hod in the Carmel Mountains south of Haifa, had to be evacuated. *Sebastian Scheiner/Keystone*

SERRANO, BRAZIL, JANUARY 2011

More than sixty people were killed in Brazil in landslides after three days of incessant rain in January 2010. A year later, a single day of heavy rainfall brought as much rain as is usually recorded in one month. Steep slopes in the hilly area north of Rio de Janeiro became unstable, hillside slums and entire neighborhoods of mountain villages tumbled into the depths below, killing more than 700 people in mudslides. Soldiers recover the body of a victim in Nova Friburgo. *Felipe Dana/AP/Keystone*

THE CONSEQUENCES OF C

RISKS AN
VULNERA

Christian Rentsch

LIMATE CHANGE

ND

ABILITIES

Everyone on our planet will have to live with the consequences of climate change in coming decades. The average global temperature has already risen by about 0.7 degrees Celsius (1.3°F) in the past hundred years, and if projections prove to be correct, temperatures will continue to rise in the coming decades and far beyond the end of this century. The IPCC and a vast majority of climate scientists project that Earth's average temperature will rise by at least another 1.1 degrees Celsius (2°F) by 2100, and more likely by 1.8 degrees (3.2°F), or even more, depending on what measures humankind takes to reduce greenhouse gas emissions.

It is impossible to predict exactly to what extent climate change will affect the rise of sea levels, the expansion of arid and desert areas, the retreat of glaciers, or other phenomena. There are several inherent reasons for this lack of predictability. Firstly, researchers have to rely on data from comparisons with past changes in climate, or on data series that are too short to provide reliable information for long-term projections. Secondly, some climate processes are not at present fully understood. But the largest factor of uncertainty is and will remain humankind. Human behavior, human activities, and human decisions will be crucial in determining the planet's temperature, the extent of sea-level rise, and the course of other factors affecting Earth's climate.

A working group commissioned by the Intergovernmental Panel on Climate Change compiled a thousand-page contribution for IPCC's Fourth Assessment Report. Entitled *Impacts, Adaptation and Vulnerability,* this section of the report screened, reviewed, and summarized several thousand studies by hundreds of independent research groups at leading universities and institutions around the world. In the course of its work, the working group made a few errors, and the findings of some studies with inadequate

scientific validation found their way into the report. This mishap, followed by an unfortunate reaction on the part of the IPCC when it became public, led to allegations of unprofessionalism directed at the IPCC. But critics failed to admit that they hadn't been able to find any other errors in the report despite a detailed search. Altogether, this section of the IPCC report on the consequences of climate change in fact provides a very reliable picture of current and scientifically sound research.

Recognizing inherent uncertainties, the IPCC made only a few very conservative and broadly worded statements on expected consequences—not in the report itself, but in its summary conclusions. It had good reason to do so—many individual case studies showed that expected changes in climate will have varying consequences in different climate zones, world regions, and individual countries. The rise in sea levels will have a completely different effect on the inhabitants of small South Pacific islands than on large coastal cities in the same region. The expansion of arid zones will affect subsistence farmers and herders in the Sahel in a completely different way than it will affect people in the Midwest of the United States or in California. In other words, the consequences of climate change for people in affected countries, and the vulnerability of each society, largely depend on the financial and technical resources and options at their disposal to protect them from negative outcomes, or to help them adapt.

There is only one thing we know with absolute certainty. Climate change is already happening—and this is just the beginning. Extreme weather conditions and the minor and major natural disasters we are experiencing today are perhaps just the first harbingers of change that will continue to unfold in coming decades.

Tropical storms form over oceans warmer than 27 degrees Celsius (80°F). Whether their frequency or intensity will increase in a warmer climate is still unclear. Cyclone over Australia, 2011. *NASA/Corbis*

Economic losses from extreme weather events increase when prosperity goes up. The path of destruction after a tornado hits Dyer County, Tennessee, USA. *John L. Focht/Keystone*

Meteorological conditions with unusual pressure patterns are responsible for repeated record rainfalls and floods in central parts of Europe. Dresden, Germany, 2002. *Ralf Hirschberger/Keystone*

Densely populated cities in coastal areas are particularly vulnerable to hurricanes, cyclones, and flooding. New Orleans after Hurricane Katrina, 2005. *Smiley N. Pool/The Dallas Morning News*

After the worst floods in the history of Pakistan, it was months before people in Karampur could finally return to their villages. September 2010. *Amjum Naveed/Keystone*

Waiting to be rescued—a helicopter evacuates inhabitants of the village of Quba Saed Khan in the province of Sindh, Pakistan, August 2010. *Kevin Frayer/Keystone*

Cut off from the rest of the world, a family that found refuge on higher ground waits to be rescued in Pannu Aqil, Sindh province, Pakistan, August 2010. *Nadeem Khawer/Keystone*

People desperate to escape flooded areas attempt to be lifted off. District of Muzaffargarh in the province of Punjab, Pakistan, August 2010. *Adrees Latif/Reuters*

A grim outlook — meteorologists expect the incidence of sandstorms to increase because of prolonged aridity.
Riyadh, Saudi Arabia, 2009. *AFP/Guardian News & Media Ltd 2011*

Increasing drought in the western United States sparks devastating forest fires in California and Montana almost every year. Bitterroot River, Bitterroot National Forest, Montana. *John McColgan/Keystone*

Out of control — when strong winds fan forest fires, even firefighters often have no choice but to flee. Gippsland region in Victoria, Australia, 2009. *Alex Coppel/Newspix*

Snow in the desert — global warming can have unexpected consequences when high humidity leads to precipitation in normally arid zones. The Sonoran Desert, Arizona, USA. *Craig K. Lorenz/Keystone*

SEVERE WEATHER AND NATURAL DISASTERS: THE HARBINGERS OF CLIMATE CHANGE

Climate change is a gradual, inconspicuous process, and it will continue long after greenhouse gas emissions have stopped increasing or even started to decline. This time lag is caused by the oceans, which take much longer to warm up than the atmosphere does.

The effects of global warming are already noticeable today in a number of minor ways. Many changes are not very dramatic and hardly seem threatening at first. Does it matter if climate scientists say that sea levels will rise by twenty to thirty centimeters, or even a bit more, in the next one hundred years? If sea levels fluctuate anyway by at least two meters between low and high tides every day, why should we be worried about twenty to thirty centimeters? Who cares if timberlines in the Alps and American mountain ranges shift upward by 150 meters? Even serious environmental disasters—droughts, floods, landslides, hurricanes, and storm surges—haven't they always happened? Aren't they just the outcome of freak weather conditions?

Climate change is more than just the sum of the changes taking place independently of each other in different parts of the world. It is the transformation of an entire system, the convergence of numerous shifts that mutually influence each other, overlap, and sometimes intensify each other's effects in feedback mechanisms. Desert areas that spread, and alpine glaciers that melt, are the outcome of several simultaneous changes such as the rise in temperature, altered wind systems, and new precipitation patterns. There are other factors too, such as human interference in the natural environment and vegetation cover. This includes the large-scale clearing of rainforests, the draining of wetlands, new agricultural land uses, and much more. Together all these changes generate a dynamic that isn't necessarily linear; in fact in some cases it can lead to a complete reversal in developments. A good example is the gradual melting of large alpine glaciers, bringing high water levels and flooding to lower river basins every spring until glaciers recede beyond a certain point; then river water levels suddenly dwindle, not only in the spring.

Another example of the non-linear behavior of the Earth system is the impact of CO_2 on plants. It has been known for more than two centuries that an increase in atmospheric carbon dioxide levels is favorable for many plants, enabling them to grow faster and better because more carbon is available for their cell development. Laboratory and field experiments have shown that carbon dioxide fertilization can boost cell growth by up to 40 percent. This even holds true for CO_2 concentrations well beyond current atmospheric levels. But if general conditions such as temperature, soil nitrogen levels, or soil moisture deteriorate, then a high level of CO_2 in air can suddenly have a negative effect on plant growth—and the system overturns. Increasing soil salinity, lack of nutrients, and water scarcity are already causing crop yields to diminish substantially in many parts of the world. Some scientists suspect that by the middle of this century, even the lush rainforests that play a vital role in reducing greenhouse gases may start to die off faster than they can grow back.

Researchers expect climate change to lead to increasingly greater environmental disasters, mostly because of more frequent and more severe extreme weather events. Numerous studies predict that these disasters will threaten climatically vulnerable regions such as monsoon areas in Asia and the hurricane-ridden Caribbean, as well as the temperate zones of Europe and North America.

Even relatively small climate shifts can lead to a marked rise in the number of extreme weather events. One example is the Vb (five B) weather pattern in Europe, a low-level vortex that carries large quantities of moisture from the western Mediterranean over Austria to countries in Central and Eastern Europe, bringing several days of heavy downpours. This uncommon but not entirely rare weather system used to occur during the cold months between fall and spring, and was responsible for large amounts of snow or moderate flooding.

For several years now, the Vb pattern has been appearing with increasing frequency during the summer months, with dramatic consequences. Since air absorbs much more water in summer than during cooler seasons, Vb summer rains tend to be much heavier and more violent. This has led to much more severe storms in Eastern and Central Europe in recent years. The Oder flood that hit Poland, the Czech Republic, and Germany in July and August of 1997, caused several large dams to burst and was responsible for widespread flooding. In Poland and the Czech Republic, rainstorms led to the deaths of 114 people and caused over 4 billion euros in damages. In August 2002, flooding along the Elbe and numerous tributaries in large parts of eastern Germany and the Czech Republic reached all-time highs, which weather experts had never thought possible. Large tracts of land with more than thirty cities and towns, including large parts of the historic centers of Dresden and Prague, and more than five hundred small communities, were under water for an entire week. The floods took their toll, affecting roughly 1.3 million people and leaving 38 dead. Damages amounted to over 18 billion euros.

Meteorologists are correct to point out that such catastrophic events are due to an unusual combination of abnormal weather conditions and are not necessarily a direct consequence of climate change. On the other hand, a marked increase in extreme weather conditions in many parts of the world, leading to a growing number of record-breaking weather events, does suggest that they are indeed signs of a changing climate.

It might be surprising to learn that the outcomes of climate change aren't necessarily similar in nature. While prolonged summer rainfalls have increased in Central Europe, for example, so have heat waves with extremely high, previously unrecorded temperatures. Such seemingly contradictory weather patterns can also be expected in winter. Climate scientists predict an overall decline in extremely cold weather periods, but at the same time, single episodes of violent winter storms with ice-cold temperatures will occur more frequently.

Interpretations of the significance of extreme weather events will certainly vary, but projections even in the very cautiously worded 2007 IPCC report leave little doubt. The report considers a "higher frequency and intensity of heat waves" and "heavy precipitation events" to be "very likely," indicating a probability of 90 to 99 percent. Furthermore, it predicts that an increase in "intense tropical cyclones" and "extreme high sea level," resulting in the flooding of low-lying coastal areas, is "likely," indicating a probability of 66 to 90 percent. These and other carefully worded statements in the IPCC report imply uncomfortable statistical truths—weather events that are considered extreme today will become so frequent as to be considered almost normal in the future. At the same time, the most severe weather events will become even fiercer in the future.

What is particularly worrisome is that the record-breaking or "hundred-year" events we observe and measure today are already occurring on an almost yearly basis, and this is just the beginning. Global average temperatures in the past hundred years have risen by only about 0.7 degrees Celsius (1.3°F), but even moderate global warming scenarios in the IPCC report predict that warming will continue and temperatures will rise by another 1 to 2 degrees (1.8–3.6°F) by the end of this century.

This increase could actually be as much as 3 or 4 degrees (5.4–7.2°F) unless more stringent measures to mitigate climate change are adopted. There is consent now that negotiations should ensure that global warming does not exceed the two-degree target; in other words, not rise more than 2 degrees Celsius (3.6°F) above preindustrial temperatures. Today many climate scientists doubt that we will manage to meet this target. They also question whether it is actually low enough to prevent dramatic or irreversible consequences of climate change.

A SHORT HISTORY OF THE TWO-DEGREE TARGET

In December 2010, at the Climate Change Conference in Cancún, 194 parties to the United Nations Framework Convention on Climate Change (UN-FCCC) agreed on the two-degree target. The agreement states that the average global surface temperature is not to rise beyond 2 degrees Celsius (3.6°F) above preindustrial levels, and substantiates Article 2 of the UN-FCCC, whose objective is to "prevent dangerous anthropogenic interference with the climate system."

The two-degree target has a long history. As early as 1996, the European Union Council of Environment Ministers, on the strength of scientific literature, declared in rather difficult but consummate diplomatic language that "global average temperatures should not exceed 2 degrees above pre-industrial level and that therefore concentration levels lower than 550 ppm CO_2 should guide global limitation and reduction efforts." At that time, researchers still assumed that it would be possible to meet the two-degree target if the level of atmospheric CO_2 stabilized at 550 ppm, whereas today they calculate that CO_2 levels should not exceed 450 ppm. Indeed, the most recent estimates indicate that the maximum concentration should be 400 ppm in order to have at least a two in three chance of keeping the average global temperature below the two-degree limit.

From 1996 onward, the two-degree target was repeatedly discussed at national and international climate negotiations until it was finally adopted as a benchmark. In July 2009, at the G8 (+5) summit in L'Aquila, Italy, the eight most important industrial nations and five emerging economies included it in their declaration on climate change. It was "noted" at the UN Climate Change Conference in Copenhagen in November 2009, and at the climate summit in Cancún in December 2010 it was finally officially recognized and declared a binding goal for future climate negotiations. →p.500

The most startling side of this long story about the two-degree target is that not a single document has ever explained why "dangerous anthropogenic interference with the climate system" starts at 2 degrees Celsius (3.6°F) and not at 1.5 degrees (2.7°F) or even 3 degrees (5.4°F).

Researchers Carlo and Julia Jaeger at the Potsdam Institute for Climate Impact Research addressed this question, otherwise rarely subject to public debate, in their paper, "Three Views of Two Degrees," and came to a surprising conclusion: "The 2° Celsius target has emerged nearly by chance, and it has evolved in a somewhat contradictory fashion: policy makers have treated it as a scientific result, and scientists as a political issue."

The two-degree limit first appeared in a discussion paper by American economist William D. Nordhaus in 1977. In a graph predicting the development of average global temperature due to an increase in CO_2 levels, the two-degree limit marked the "estimated [temperature] maximum experienced over [the] last 100,000 years." In a commentary accompanying this graph, Nordhaus wrote that this figure could serve as a first approximation for a critical limit because "it seems reasonable to argue that the climatic effects of carbon dioxide should be kept within the normal range of long-term climatic variation.... If there were global temperatures more than 2 or 3 degrees above the current average temperature, this would take the climate outside of the range of observations which have been made over the last several hundred thousand years." In other words, Nordhaus's paper didn't give any evidence that exceeding this two-degree limit could have critical consequences for the climate.

Nordhaus's paper did not draw much attention and the two-degree limit was not mentioned in the IPCC assessment reports or any other scientific publications. Not until 1990 did it re-emerge in a report published by the Advisory Group on Greenhouse Gases (AGGG), a joint advisory committee of the International Council of Science Unions, the World Meteorological Organization, and the UN Environment Programme. The two-degree limit was now declared the "upper limit beyond which the risks of grave damage to ecosystems, and non-linear responses, are expected to increase rapidly." The report does not give any indication of how the advisory group reached this conclusion.

In 1995, the German Advisory Council on Global Change, established by the German government, adopted this limit with similar reasoning, and in 1996 the EU Council of Environment Ministers also accepted it. Since then it has been used as a target for climate policy at nearly all international environmental and climate change conferences.

But Carlo and Julia Jaeger illustrate that the reasoning behind the two-degree limit has turned a few corners. While Nordhaus himself did not hold forth on any potential risks, experts in the 1990s pointed to the existence of so-called tipping elements, natural mechanisms that might trigger non-linear and irreversible processes once changes in the climate reached a critical point, with consequences that would be nearly impossible to predict. They referred, for example, to the thermohaline circulation, and especially the northward extension of the Gulf Stream, which is responsible for temperate climate in Northern Europe. If this circulation reached a certain tipping point and abruptly came to a halt, the climate in Europe would cool substantially within a few years. However, more recent studies show that such an abrupt event would probably happen only at much higher temperatures and consequences would be less severe than originally surmised. According to the researchers from Potsdam, this illustrates that the thermohaline circulation cannot be used to justify the two-degree limit.

Yet Jaeger and Jaeger also point to studies demonstrating that a rise in global average temperature of only 1.5 degrees Celsius (2.7°F) acutely increases the risk of severe or even irreversible outcomes, for instance for global water supplies. In another example, if temperatures in polar regions go up much higher and more rapidly than the global average, sea levels will rise far more than currently estimated.

The researchers therefore conclude that from a scientific point of view, the two-degree limit is not suitable "as a threshold separating a domain of safety from one of catastrophe, and as an optimal strategy balancing costs and benefits." They add that the two-degree limit is above all a political target, a kind of "focal point in a coordination game, where a multitude of actors need to find a new coordination equilibrium in the face of climate risks.... The focal point may then be redefined on the basis of experience."

In other words, it may be possible to scientifically justify the two-degree limit, but this value has not been proven scientifically valid for curbing risks. Mostly it is the expression of a certain attitude of responsibility for protecting the climate and the world. Reto Knutti, climate scientist at the Institute for Atmospheric and Climate Science in Zurich, writes, "There is no clearly defined limit that separates dangerous climate change from harmless climate change. It is difficult for instance to assess the damage caused when an animal species becomes extinct. Even a large amount of money cannot reverse such a loss. A maximum temperature increase of 2 degrees is thus a subjectively chosen threshold.... The bold two-degree target is perhaps the best we can hope for and the worst we can tolerate."

ARCTIC WINNERS

Our planet is getting warmer everywhere, even in places where it used to be too cold for year-round human presence. A new frontier for economic interests is opening up—ice-free shipping routes, new agricultural lands, and access to resources buried under the eternal ice now hold promise. There is concern that these activities will put additional pressure on the fragile ecosystems of the Arctic.

Klaus Lanz

Until recently, the Arctic seas to the north of the Asian continent—the Kara, Laptev, East Siberian, and Chukchi Seas—attracted little international attention. Located far north of the Arctic Circle, they are barely accessible and of little, if any, economic interest. These areas experience extremely low temperatures during most of the year, their waters remain frozen for months, and their coasts are virtually uninhabited.

But in the past few decades, global warming has also affected the north polar region and its impact is quite noticeable. Temperatures in Arctic latitudes are rising two to three times faster than in temperate latitudes. Ice is steadily melting, and the summer sea ice cover is diminishing. Ice becomes thinner and more brittle every year, and in 2007 the ice cover receded more than it ever had before. In the summer of 2009, two German merchant ships succeeded in sailing through the Northern Sea Route (Northeast Passage) from Norway to Vladivostok without the assistance of an icebreaker.

These altered conditions have put the Arctic seas increasingly into the focus of economic interests. The Chinese government also took notice of the pioneer voyage of the German ships. Since nearly half of Chinese foreign trade is carried by ship, there is growing interest in a Northeast Passage that is open for several months during the summer. The northern shipping route from Shanghai to Hamburg across the Bering Sea and past northern Norway is 6,400 kilometers shorter than the route through the Malacca Strait and the Suez Canal. There are also safety aspects to consider—

Most of the Arctic is still a region of unspoiled nature. That could soon change as geologists suspect that huge oil and gas reserves are under the Arctic's ice. *Jean Gaumy/Magnum Photos*

insurance premiums for shipping have gone up tenfold because of pirate attacks off the coast of Somalia in 2009.

The sea route from China to the east coast of the United States is also substantially shorter through the Arctic. Nevertheless, it still remains to be seen whether new polar transportation routes between Asia, Europe, and North America will ever be used on a commercial scale. There are few ports along the coasts of the northern seas, and infrastructure is practically nonexistent. In spite of a warmer climate, extreme weather and adverse sea conditions would continue to make shipping hazardous. Even though Arctic sea routes will be mostly free of ice during future summers, ships would still have to deal with floating ice. If the Greenland ice sheet continues to melt, huge icebergs could slow down freighters and force them to take detours. Furthermore, not all cargo ships can pass through the Northern Sea Route, since waters in the Bering Sea, for instance, are not always deep enough for the largest container ships.

Russia also has an interest in promoting shipping along its north coast. In the summer of 2010, a double-walled, ice-class cargo ship carried 90,000 tons of Russian gas condensate through the Bering

Strait to the Far East. At least eight such voyages a year are planned for the future. Shipping provides the Siberian economy with new distribution channels for its rich mineral resources and especially for timber from its huge boreal forests. Conditions in the high north have already changed significantly because of climate change. During the Cold War, the Soviet Union managed to maintain its military bases on the north coast only with the help of a huge fleet of icebreakers. But in future summers, shipping routes along the north coast of Russia are expected to be entirely free of ice for several weeks.

It is difficult to predict how accessibility to the Arctic Ocean and its coasts will develop from year to year. Ice formation in winter and ice melt in summer depend not only on average global temperatures, but also on cloud cover, winds, and warm sea currents. But in all probability, the sea ice cover will continue to recede, both in thickness and extension.

Regardless of precisely when the Arctic Ocean will be free of ice, be it as early as 2030, or not until 2050, as predicted by the head of the Russian Meteorological Center in 2010, one thing is certain: ocean carriers around the world can hope for significantly improved conditions for shipping in the Arctic Ocean in the near future.

AGRICULTURAL DEVELOPMENT IN THE ARCTIC

Projections on the future of agriculture in northern latitudes are even more difficult to make than those for ice conditions. The Arctic Climate Impact Assessment (ACIA) indicates that climate change could enable the cultivation of potatoes in virtually all of Scandinavia by the end of the twenty-first century. The northern temperature limit for grain cultivation could shift to Fairbanks in Alaska, and Novosibirsk and beyond. It is even conceivable that sunflowers could grow in Tromsø and Reykjavik by 2090, depending on the rise in temperature.

But these projections are only half the picture. Plant growth and crop yields depend on many factors, not just average temperatures. We can expect the number of days when plants are able to grow in Alaska, Siberia, Iceland, or northern Scandinavia to increase appreciably by the end of this century. But agriculture will not flourish everywhere within the Arctic circle, primarily because of a lack of water. And even in a warmer future, many perennial crops won't be able to withstand bitter winter frosts.

Potatoes, rye, and barley have been cultivated in Alaska only in small quantities so far. The expected rise in temperature could make extensive farming in northern regions of Alaska interesting for the agricultural industry. But according to forecasts in the ACIA, the boreal region above the 50th parallel north would require artificial irrigation to enable farming. Climate scientists expect global warming to lead to some more precipitation in this dry region, but given that evaporation will also increase, rainfall alone would be unlikely to provide enough water for agricultural cultivation in most years. On the mountainous west coast, where precipitation is more abundant, there is a lack of suitable soils and level, arable land. Apparently the U.S. Department of Agriculture is not convinced of Alaska's future agricultural potential since it has taken no steps so far to encourage farming in Arctic regions.

The provinces of northern Canada, Yukon and Northwest Territories, are very mountainous. There is potential for agricultural activity only in a few valleys. Here, higher temperatures are not expected to improve farming conditions because of the region's lack of fertile, nutrient-rich soil. In addition, Yukon's precipitation levels are so low that grass steppes dominate the landscape instead of forests. In subarctic Canada, agriculture is thus likely to always remain a marginal activity.

The same is true for the very far north of Europe, where the probability of climate change improving the potential for agricultural cultivation is low. Rising temperatures will prolong the growing season and extend the northern limit for the cultivation of grain, but otherwise, conditions are unfavorable. In Finland, soils are generally very shallow and infertile, making them unsuitable for cultivating grains and vegetables. In Norway, the Gulf Stream causes summers to be relatively warm, but also cloudy and rainy, which lowers crop yields. Agriculture in Norway will prosper only if climate change brings both warmer and sunnier summers—which current models cannot predict with certainty.

Practically all climate-induced developments in high-latitude agricultural cultivation will depend more on rainfall than on temperature. Potential gains because of longer growing seasons are offset by losses from a lack of rain and higher evaporation rates. From today's perspective, we can say that anyone hoping for a northward expansion of farmlands and a subsequent increase in food production in subarctic regions will most likely be disappointed.

Tiksi, the Arctic port of the Sakha Republic (formerly Yakutia) in Eastern Siberia, hopes that shipping, trade, and natural gas extraction in the far north will bring economic stimulus. *Pavel Kolínský*

ACCESS TO OIL AND GAS

If there are winners benefiting from the Arctic thaw, they will undoubtedly be the oil and gas industries. Huge fossil fuel resources will become accessible if the Arctic Ocean remains free of ice for a few months each year. The United States Geological Survey (USGS) estimates that 13 percent of the undiscovered oil reserves worldwide and as much as 30 percent of the undiscovered gas resources are to be found in the Arctic region, most of it offshore.

The Arctic states, Russia, Canada, the United States, Norway, and Denmark, are eager to divide these natural resources among themselves and exploit them as soon as possible. But there is one problem: no overall agreement has yet been reached on the exact demarcation of borders in the Arctic Ocean. Norway and Russia signed an agreement in September 2010 settling a forty-year dispute over their border in the Barents Sea between the islands of Novaya Zemlya and Svalbord. On the very same day, they began talks on the exploitation of oil and gas resources in this marine area, closed for decades because of the conflict.

New perspectives — China's sea routes to Rotterdam and New York will be shorter if Arctic ice continues to melt.
Hao Xiaoguang / Chinese Arctic and Antarctic Administration

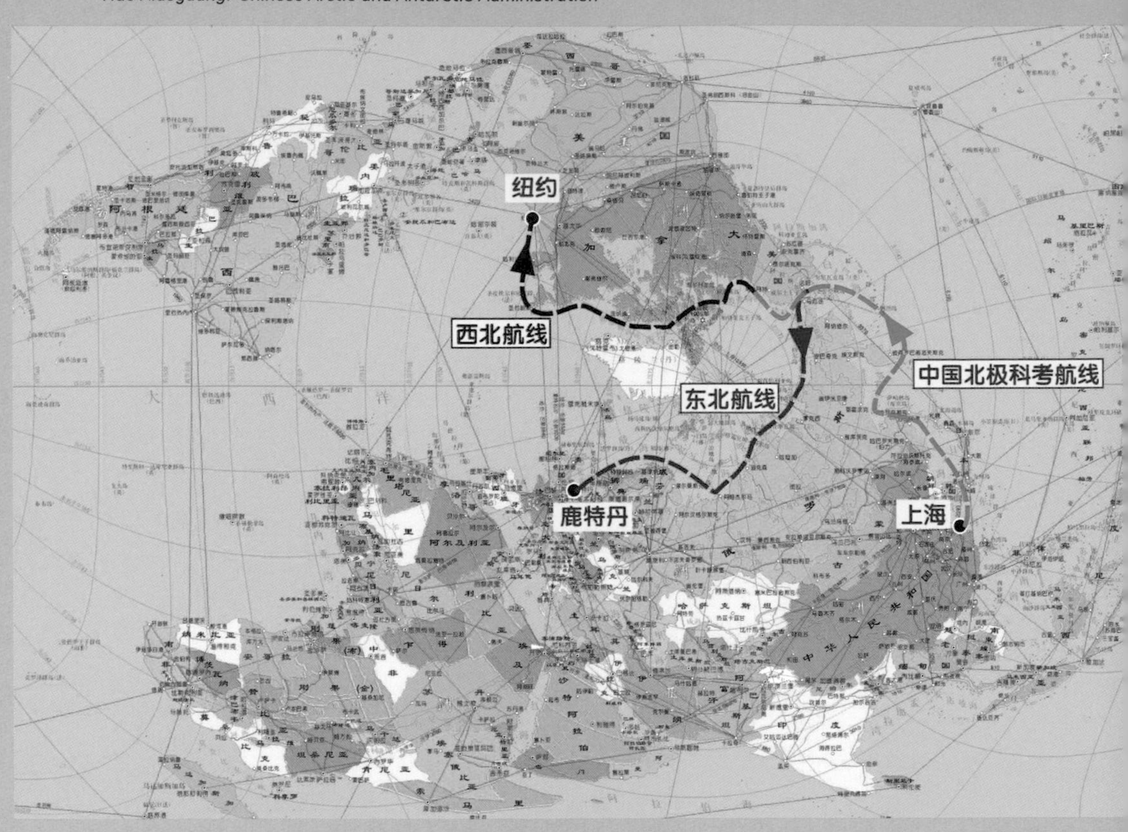

Large energy corporations would like to lay their hands on Arctic reserves, and the sooner the better. For many years, Russia's Gazprom company, together with Norwegian and French partners, has been attempting to exploit a gas field west of Novaya Zemlya called Shtokman, located at a depth of over 300 meters. Offshore extraction is difficult, hazardous, and expensive because of the extreme Arctic environment. In 2010, the Shtokman project was put off for the time being because of falling gas prices. On the other side of the Arctic Ocean, in the Beaufort Sea north of Alaska, Shell Oil applied to American authorities in 2010 for approval to drill an exploratory well, with others to follow. Norway's Statoil has already put into operation the first gas field in the Barents Sea called Snøhvit ("Snow White"). Gas condensate has been flowing since 2007 via a 143-kilometer-long underwater pipeline to Melkøya Island for processing.

It remains to be seen how peacefully the further exploitation of Arctic energy reserves will proceed. The United States and Russia, but also Denmark, Norway, and Canada, are preparing to strengthen their presence in the Arctic. U.S. military officials have called for additional icebreakers to secure American interests in the region.

Even China, which does not border the Arctic, has recognized the region's strategic importance and is monitoring the situation carefully.

It is clear that energy corporations stand to profit the most from climate change in the Arctic. The Arctic Ocean is still almost untouched by human activity and its environment is largely intact compared to that of other oceans. But now, with the diminishing of frost and ice, humankind has begun to stake its claims to the region. Unless the entire Arctic Ocean is placed under protection, it will become a gigantic development area for fossil fuels. If exploitation goes ahead, the unique environment of this last virtually untouched ocean on Earth will be lost. And as more fossil fuels become available for burning, climate change will advance even more rapidly, bringing with it insufferable loss of another kind.

A DIFFERENT WORLD FOR BILLIONS OF PEOPLE

At this point, no one knows if we will succeed in adopting and implementing effective measures to avert "dangerous anthropogenic interference with the climate system," the declared objective of the Framework Convention on Climate Change. It is not even clear whether the two-degree target adopted in Cancún will be low enough to prevent serious or even irreversible consequences of climate change. Recent research indicates that such changes will already occur when global warming exceeds 1.5 degrees Celsius (2.7°F).

A study group commissioned by the IPCC reviewed and analyzed numerous single case studies. They published their findings in a contribution to the 2007 IPCC report of more than 1,000 pages, entitled *Impacts, Adaptation and Vulnerability.* In their conclusion, the *Synthesis Report,* researchers made six very general statements:

- An increase in warmer and hotter days and nights is virtually certain over most land areas.
- An increase in the frequency of warm spells and heat waves over most land areas is very likely (indicating a probability of over 90 percent).
- An increase in the frequency of heavy precipitation over most areas is very likely.
- An increase in areas affected by drought is likely (indicating a probability of over 66 percent).
- An increase in intense tropical cyclone activity is likely.
- The increased incidence of extreme high sea levels is likely. (Tsunamis triggered by earthquakes are explicitly excluded.)

What do these sober IPCC statements mean? First of all, climate change affects us all, and it affects, in varying degrees, all latitudes, world regions and areas of life, urban and rural alike. Even though these statements are very general, climate change has vastly different consequences and effects at the regional level.

Moreover, countries and their populations are affected in different ways. The industrial nations of the North, because of their affluence, will be better able to protect themselves against inevitable consequences than the large emerging economies in Asia and Latin America with their huge and still mostly very poor populations. The poorest developing countries in Africa will barely be able to protect themselves from the effects of climate change without massive aid efforts. Furthermore, the consequences of climate disasters will keep them trapped in a vicious circle of underdevelopment. This means that for one to two billion people already living under very precarious conditions today, even the mere access to food and water will be difficult, poverty will increase, and hygienic conditions will continue to deteriorate; in any case, infant mortality and the risk of epidemics will certainly not diminish.

The industrial nations of the North will not be spared either. It is very likely that they will experience more frequent and severe natural disasters, with floods and heat waves, storm damage and crop failures. Vulnerable technological infrastructure will be disrupted by power outages and breakdowns in transportation, which could lead to substantial economic losses.

Even if we leave natural disasters aside, the world will change in essential ways. Evidence of these changes is already noticeable now, although the rise in temperature of 0.7 degrees Celsius (1.3°F) in the past hundred years is still distinctly below expected increases according to best-case scenarios. The amount and seasonal distribution of precipitation have already markedly changed in many parts of the world. Timberlines in many mountain regions have advanced upward by one hundred meters and more, and in subarctic regions they have shifted hundreds of kilometers toward the north. The onset of (climatic) spring now occurs, with regional differences, up to two weeks earlier, and autumn starts ten days or more later.

Extended plant growth periods caused by a warmer climate can also have positive effects, particularly in the Northern Hemisphere, in Northern Europe, North America, and Russia, since cultivated areas can be expanded and the growing period has lengthened. This is only useful, however, if altered flowering and ripening periods, overfertilization, or soil degradation don't interfere. In other parts of the world, for instance in India, Eastern Africa, or Australia, the tendency is for growing periods to shorten because of the shift in dry and rainy seasons, making it impossible to cultivate plants that need more time to ripen. Experts estimate that in Africa's arid and semi-arid zones crop yields could diminish by up to 50 percent as early as 2020.

Changes in atmospheric circulation patterns over the tropics, responsible for a shift in climate zones, have a major impact →p.136. Stronger heating around the equator, the Intertropical Convergence Zone, causes the Hadley cells to expand farther north and south, thus pushing adjacent climate zones to higher latitudes.

Meteorological studies show that climate zones are shifting faster than earlier climate models had predicted, having reached a dimension today that earlier models had projected for the end of the century. Some studies assume that climate zones adjacent to the tropics will shift about 300 kilometers poleward. This development could definitely have a positive effect on subtropical arid and semiarid areas such as southern parts of the Sahel region, since they would receive more rainfall in the future than they do now. On the other hand, precipitation is expected to further decrease in bordering regions such as the Mediterranean and the Middle East, the American Southwest and northern Mexico, in large reaches of southern Africa and Australia, and some parts of South America—indeed, some models predict up to 50 percent less rain in their worst-case scenarios.

A 2007 study by the American Academy of Sciences predicted that by the end of the century, some 48 percent of the world's regions will be subject to entirely different climatic conditions. It is even possible that in some parts of the tropics and subtropics, as well as in various mountain regions and at the poles, completely new climate patterns will form that Earth has never seen before.

A shift in climate zones poses a special threat to ecosystems, especially when they have already been disrupted or damaged by other factors, mainly human interference. Increasing population growth, deforestation, the draining of wetlands, the channeling of rivers, and changes in land use for agriculture, urbanization, industry, transportation infrastructure, and tourism all take their toll. How well ecosystems can adapt to climate change also depends on how rapidly the change occurs. Scientists estimate that most ecosystems can cope with a maximum rise in temperature of 0.1 degrees Celsius (0.2°F) per decade, and that a rise of 0.3 degrees (0.5°F) per decade would be far too much. Current estimates predict a rise in temperature of 0.2 degrees (0.4°F) per decade.

Biodiversity is especially at risk. Particularly sensitive ecosystems such as coral reefs are already showing signs of stress. Ecologists estimate that if temperatures were to rise by 0.8 to 1.7 degrees Celsius (1.4–3.1°F) by 2050, 18 percent of a long list of surveyed plants could become extinct. If the temperature rose by 1.8 to 2 degrees (3.2–3.6°F), that figure could go as high as 24 percent. According to the International Union for Conservation of Nature's (IUCN) Red List of Threatened Species, 20 percent of mammals, 12 percent of bird species, and 29 percent of amphibians are threatened with extinction.

The executive summary of the 2007 IPCC report states that: "During the course of this century the resilience of many ecosystems ... is likely to be exceeded by an unprecedented combination of change in climate, associated disturbances (e.g. flooding, drought, wildfire, insects, ocean acidification) and in other global change drivers (especially land-use change, pollution, and over-exploitation of resources), if greenhouse gas emissions and other changes continue at or above current rates."

Even without severe natural disasters caused by extreme weather events, future generations will most likely live in a world very different from the one we know today.

NEARLY HALF THE WORLD SUFFERS FROM WATER SCARCITY

The number is impressive—41 percent of Earth's surface area is drylands, amounting to 61 million square kilometers—an area twice the size of Africa. Although polar deserts and tundra regions in Canada, Greenland, and Eastern Siberia also count as drylands, the largest areas are made up of the huge deserts and steppes of North Africa, the Middle East, and Central Asia, as well as great expanses of land in the western United States, the Pacific coast of South America, Australia, and southwestern Africa.

Annual precipitation has declined in the past century in many of these areas, and drylands have expanded. The driest regions, receiving very little or no rain for periods of nine months or longer, have doubled and now make up as much as 30 percent of the planet's land surface. According to the 2007 IPCC report, these regions will continue to increase in size until the end of this century. Other regions that are becoming more and more affected include the Mediterranean, other parts of Africa, Central America, India, Pakistan, and a wide belt that stretches from New Zealand to New Guinea and up to Japan.

This is particularly dramatic because these arid and desert regions are home to more than two billion people—nearly one-third of the global population. Many of these people are already barely able to survive, especially those millions of subsistence farmers and herders who live almost exclusively on the basis of yields from their small fields and a few goats and cattle. The unrelenting expansion of dry regions therefore inevitably entails the peril of hunger and malnourishment, especially because populations in many of these regions grow at a much faster rate than in the industrialized West.

But this is not the place to examine the complex issue of food security on a global scale, or analyze the problems of the global food market and distribution, or question market mechanisms and speculation. Here we are looking only at the consequences of climate change. But as the impact is most pronounced in the most densely populated and poorest of the world's developing countries, we cannot ignore the situation of those millions of chronically undernourished human beings who simply wish to survive, but who are not able to

“Human beings are a strange species, and unique in the universe. The dedication with which they saw off the branch on which they sit places them apart from other animals. Considering the age of the planet, they are managing to destroy it in an incredibly short time. A few milliseconds on the world clock—and they have already managed to darken the sun and poison the air. The blue seas are no longer blue, and Earth’s eternal ice is melting away.”

Thomas Assheuer, editor at the German weekly newspaper *Die Zeit* participate in the free market game of supply and demand because they have no money.

The way in which people in these areas live is not a vision of the future that environmentalists use to put moral pressure on climate negotiations—for the thirty to forty million people of the Sahel region, the worst-case scenario has already been very real for about forty years. Rainfall in the Sahel, a 700-kilometer-wide belt spanning the entire African continent from east to west, has declined by nearly half since the 1970s. Pastures and croplands that previously had allowed modest agricultural activity were already suffering damage from overuse and soil erosion forty years ago. Since then, this region has turned into a desert-like landscape. Mauretania, Mali, Niger, Chad, and Sudan have been afflicted by a rapid succession of prolonged droughts and devastating famines. Even years with higher-than-average rainfall, occurring sporadically since the 1990s, cannot reverse the process of desertification. This disaster in the Sahel region is only one example.

No one can predict what climate change will mean for Africa. The continent's population is expected to double by the middle of the century, growing to two billion. Barren lands and deserts already occupy two-thirds of its area. The world's poorest continent is ravaged by wars and disease, and its megacities are overflowing with people who have migrated from rural homelands in the illusory hope of finding work in cities so they can somehow survive.

To this day, researchers have not been able to fully determine the degree to which growing aridity is an outcome of anthropogenic climate warming. Many climate scientists believed until recently that desertification on the fringes of the Sahara and Gobi deserts was caused mainly by overgrazing, overcultivation, and the excessive clearing of vegetation. According to this hypothesis, these activities led to permanent damage of the vegetation cover, and the already meager soil was no longer able to absorb enough water during the rainy season. Sparse vegetation no longer released moisture, causing rainfall to decrease even more—a vicious circle.

However, recent studies and improved climate models show it is highly probable that climate warming does in fact play a rather significant role in the expansion of drylands, with the overuse of soil merely intensifying the effect. The amount of rain is regulated primarily by ocean evaporation and the strength of winds that carry moisture-laden air to the interior of a continent. Wind force in turn depends on the difference in temperature between oceans and land masses. This difference has declined considerably since the 1970s and 1980s, coinciding with the beginning of the great Sahel droughts, because the Indian Ocean has warmed up more than average.

Several models predict that ocean warming and other climate effects, such as a shift in climate zones and a more intense El Niño phenomenon, could cause rainfall in many semi-arid, subtropical regions to decline even further in coming decades.

But crop yields will also be decimated in areas with more abundant rainfall if rainy seasons shift by several weeks, or rains become increasingly irregular. In the past few years, farmers in several central African countries have been able to grow only one crop instead of two a year. A study by the International Institute for Applied Systems Analysis (IIASA), commissioned by the UN Food and Agriculture Organization (FAO), predicts that drylands in Africa could expand by another 5 to 8 percent by 2080, and grain production further decrease by about 20 percent.

This doesn't necessarily mean that global agricultural production will decrease in the coming decades. The IIASA study concludes that warmer temperatures, increasing rainfall, and an expansion of farmlands in the United States and Northern and Central Europe could more than compensate for losses in Africa. But this is hotly debated. Many scientists find fault with the IIASA prediction because it doesn't take into account seasonal extremes or secondary environmental effects such as polluted water, overfertilization, land exhaustion, and the lowering of groundwater tables. For years now, factors such as these have led to permanent crop losses in some of the world's most important grain-producing regions.

A study by American researchers David Battisti and Rosamond Naylor on selected European countries suggests that even apparent climate "winners" in temperate zones may face substantial losses in agricultural productivity toward the end of the century. They based their calculations on seasonal temperatures, more important for plant growth and maturity, instead of on average yearly temperatures. Comparing the distribution of summer temperatures in the last century to those projected in several independent future climate models, they were able to estimate potential future crop losses.

Several things became clear. If, as expected, global warming causes average summer temperatures in Western Europe to rise by around 3 degrees Celsius (5.4°F) (slightly more than the average global increase), nearly every second summer in Europe will display weather conditions similar to the 2003 summer heat wave, with record temperatures of 40 degrees Celsius (104°F) and higher. In the hottest summers, temperatures are expected to climb to unprecedented levels. The year 2003 is an indication of what such summers could have in store for us—corn yields in Italy dropped by 36 percent, in France by 30 percent; wheat yields fell by 21 percent and those of fruit by 25 percent. Other countries, including Spain and Germany, recorded similar crop losses. High temperatures and the lack of water also directly affected humans and animals. Livestock showed clear symptoms of heat stress and the use of water had to be rationed in many places. At least 52,000 people died in France and Italy as an indirect result of the heat.

The Battisti/Naylor study states: "By the end of the century, it is very likely (greater than 90 percent chance) that a large proportion of tropical Asia and Africa will experience unprecedented seasonal average temperature, as will parts of South, Central, and North America and the Middle East." In temperate latitudes this development is predicted to be "likely." And: "It will be extremely difficult to balance food deficits in one part of the world with food surpluses in another, unless major adaptation investments are made soon to develop crop varieties that are tolerant to heat and heat-induced water stress and irrigation systems suitable for diverse agroecosystems."

A research team headed by American geophysicist Richard Seager of the Lamont-Doherty Earth Observatory of Columbia University in Palisades (New York State) has published a study forecasting developments in the American Southwest. The team's prediction is that extended droughts of the kind that used to occur only every twenty to thirty years, and which always led to major crop failures in Arizona and Texas, will probably become a permanent condition by the middle of the century, making the region even drier, especially in La Niña years. This is all the more threatening because the Ogallala

Aquifer, which supplies the Corn Belt of the United States with groundwater, may almost completely dry up within a few decades. Increased aridity will also affect drinking water supplies, especially in California, where water use is already drastically restricted by mandatory rationing almost every year.

Forest fires will probably become more frequent in these dry regions, in California as well as in the Mediterranean, in Southeast Asia, and especially in Australia. Subtropical zones are hardest hit because these fires often leave behind irreversible damage. The persistent lack of moisture and increasing summer temperatures prevent forests from regenerating and make reforestation very challenging. Once the vegetation is burnt, forest areas erode and turn into steppes or deserts. They are, like the African steppes of the Sahel, potentially lost for hundreds of years, even if there is increased rainfall in some years.

AUSTRALIAN WATER WOES

Catastrophic floods in southeastern Australia early in 2011 ended a period of extreme drought that had lasted for years. The two large rivers of the region, the Murray and the Darling, suffered from water scarcity, as did the local agricultural industry, which is strongly dependent on artificial irrigation. The Australian government is trying to tackle these environmental problems by reorganizing water distribution and modifying agricultural systems.

Klaus Lanz

The catchment area of Australia's two longest rivers, called the Murray-Darling Basin, occupies an area of more than one million square kilometers, approximately one-seventh of the Australian continent. Its waters come from innumerable smaller streams and rivers in the states of Queensland, New South Wales, and Victoria. The Murray river flows into the Indian Ocean in southern Australia near the port city of Adelaide. Irregular rainfall is typical for the climate of southeastern Australia. Periods of more or less abundant rain are usually followed by years of drought, as for instance from 1895 to 1902 (the Federation Drought) and during World War II.

The most recent period of drought, which finally ended in early 2011 with the onslaught of disastrous floods, was the worst and longest in human memory. Experts are certain that it was intensified by climate change. Year after year, starting in 1997, the region failed to get enough rain, and from 2001 to 2010 a state of emergency was officially declared because of severe drought conditions. The agricultural industry sustained huge losses from the lack of water. Rice farmers lost 85 percent of their revenue between 2001 and 2006, cotton yields dropped by 70 percent, and winegrowing was also severely hurt.

In the last four years of drought between 2006 and 2010, not a drop of Murray water ever reached the sea. Locks at the river's mouth had to be closed to prevent seawater from flowing upstream. Lake

Years of drought mean trouble for Australia's cattle farmers. Deep groundwater for livestock has to be pumped to the surface. Deniliquin, New South Wales. *Agence Vu/Keystone*

Alexandrina and Lake Albert, both located near the coast and normally fed by the Murray, sank to one meter below sea level. Surrounding floodplains dried up and salinity enormously increased. The unique Coorong coastal lagoon ecosystem usually gets its freshwater from these lakes, but because they were no longer replenished by water from the Murray, the Coorong also began to dry up and salinity levels rose. Plant and animal life in this coastal lagoon of international importance was severely disrupted.

In spite of the lack of rain and incipient climate change, humans themselves are mainly to blame for this ecological and human disaster. Forty percent of all Australian farms and roughly two-thirds of the country's irrigated fields are located in the Murray-Darling Basin. State governments, in charge of water policy, allowed farmers to divert excessive amounts of water for decades from the Murray-Darling system, supporting the industrial cultivation of cotton even as late as the 1980s and 1990s. Water consumption increased fivefold over the course of one hundred years. Today, 83 percent of the water is used for agricultural purposes (7.2 billion cubic meters). If losses incurred from the transportation of water are added to that figure, total agricultural water consumption amounts to as much as 96.8 percent.

Farmers continued to claim large amounts of water even during the recent drought. The agricultural industry was unwilling and unable to adapt cultivation to altered conditions. It was not until rivers and groundwater started to run dry in many places that cutbacks were made in the cultivation of rice, a very water-intensive crop. The fact that the rivers steadily deteriorated during the drought is due first and foremost to intensive farmland irrigation.

A RIVER WITHOUT WATER

In normal years, and in the absence of retaining dams and water diversions, the Murray would carry an average of 12.2 billion cubic meters into the ocean. Nearly half of that water is withdrawn from the river system, primarily for irrigation farming. Natural losses from evaporation, for instance, further reduce the flow of water into the ocean to a long-term average of only 4.7 billion cubic meters or 39 percent of its natural volume. This amount would probably suffice to sustain rivers and wetlands, and the lakes at the mouth of the Murray. But since agricultural water diversions continued unrestrained during the drought, absolutely no water reached the mouth of the river from 2006 to 2010.

Since 1900, Australians have been aware of the fact that too much water is withdrawn from the river system even in years of abundant rainfall. Yet for decades, the four states concerned were unable to keep water abstractions within sustainable limits—the interests of agricultural lobbies always prevailed. As soon as water became scarce in a dry period, government authorities built new dams to allow more water to be diverted from rivers.

A turning point in public opinion was reached when the locks at the mouth of the Murray River had to be closed for the first time from 1981 to 1983 to prevent seawater from flowing up the river. But authorities still continued to underestimate the problem. Although no further extraction rights for river water were issued as of 1997, farmers increased their consumption of groundwater, indirectly decimating water flow.

WATER TRADING

In an attempt to limit overall water abstraction, the Australian government decided to make use of free market principles. The idea is that when a commodity becomes scarce, trading it will automatically lead to more effective and sensible distribution. So the government created a water market, hoping that the scarce water resources of the Murray-Darling Basin would be reallocated to bring the most

After fifteen months of drought in 2007, farmers in Australia's Murray-Darling Basin set up survival feeding measures for their sheep and cattle on parched pasture lands. *Agence Vu/Keystone*

yield. For the past few years, farmers with low yields have been able to sell their irrigation rights to other farmers, who expect this water to generate higher profits. In 2008 and 2009, water rights for over 1 billion cubic meters were sold and temporary water allocations for another 1.7 billion cubic meters were granted to interested parties.

However—aside from rampant speculation and exorbitant prices during the drought period—this trade only led to a shift in water withdrawal and not to its reduction. The goal had been to retain more water in the rivers and natural wetlands, but turning water into a commodity did not help one bit. Therefore the government itself now wants to intervene in the water trade. It plans on spending A$3.1 billion to buy back water rights from farmers and declare them invalid so that the water balance of the area's natural systems can be restored in the future.

This enormous sum is part of a strategic plan designed by the Australian government to save the Murray and the ecosystems it feeds. A total of A$12.6 billion has been designated to optimize rural water management and the efficiency of agricultural irrigation systems by 2018. But it is already clear that much more drastic measures will be needed to secure the survival of river systems in the long run. Consequently, the Water Act of 2007 provides that the diversion of water resources in a catchment area must remain within limits safeguarding the long-term survival of aquatic ecosystems.

Another major novelty of the Water Act was to shift the responsibility for water management from the state to the national level. To that end, the Murray-Darling Basin Authority (MDBA) was created, an agency reporting directly to the Australian government. Under pressure because of the water crisis, the states finally agreed to give up their roles in managing and allocating water. It is now up to the MDBA to determine how much water can be taken from rivers without endangering their environmental function. Scientific studies indicate that a substantial reduction in diversions will be needed to keep river and estuary ecosystems alive.

Furthermore, more recent studies indicate that the lack of rain in the last ten years is related to changes in large-scale atmospheric circulation patterns. In other words, atmospheric conditions have changed to such an extent that the inhabitants of the Murray-Darling Basin will have to cope with significantly altered conditions in the future. The climate in the region has become more irregular, and the trend to year-long droughts, and rare but extreme torrential rains, could continue.

Even before the worst flooding in decades ended the record drought in the Murray-Darling Basin, the new MDBA agency had published its calculations for environmentally sustainable and responsible water diversion. Farmers will have to make do with about one-third less water on average in the future, and in some regions the reduction will be as high as 45 percent. There are also plans to significantly reduce the exploitation of groundwater. A reduction of the excessive acreage under irrigation and the transition to less water-intensive crops should help to economize water use and save the required amounts. However, the MDBA publication led to furious protests in the regions concerned since many farmers feared for their livelihoods.

The issue of water allocations has disappeared from the headlines since eastern Australia was devastated by catastrophic floods early in 2011. But it still continues to be a top priority in Australian politics. The Australian government will not be able to avoid taking drastic measures to limit water consumption, at the latest with the next drought. The agricultural industry in the Murray-Darling Basin, coddled for decades by local state governments, has become accustomed to

Farmers are forced to buy fodder during periods of drought, which cuts substantially into their profits. It is uncertain how long pasture farming can remain profitable. *Agence Vu/Keystone*

using amounts of water that are not sustainable, even in years of abundant rainfall, not to mention in times of future climate change. Climate change will force us all to make adjustments, and farmers' protests show us the kind of challenge that even a highly organized, prosperous, and democratic country faces. Australia gives us a glimpse of what the future could hold in store in other regions as well.

SHRINKING ICE MASSES, RISING SEA LEVELS

On October 17, 2009, the president of the Republic of Maldives and members of his cabinet convened underwater, wearing diving suits. The ministers signed a declaration calling on the governments of the world to reduce greenhouse gas emissions. This spectacular action was covered by all major international daily newspapers.

In fact, within just a few decades, 300,000 inhabitants on the world's smallest island state risk losing their home to the Indian Ocean. Even if the sea level does not actually rise by one meter or more, as some recent studies suggest, but only by the thirty to forty centimeters projected in the 2007 IPCC report, several hundred thousand people will very likely lose their homes and their livelihoods, and be forced to move to other countries in search of a new future.

The rise in sea level is not the only threat to these islands. The increasing frequency and intensity of extreme weather conditions, violent tropical storms, and flood surges are just as dramatic. They destroy the coral reefs so important to the survival of atolls, and uproot coconut plantations near the coast—often the only export product from these islands. They contaminate the islands' already scant supply of freshwater with salty seawater and erode and wash away the sandy beaches that are so vital for the tourist trade.

But if the sea level actually rose more than predicted by the IPCC report, it would be virtually impossible to save numerous islands in the Indian Ocean and South Pacific, and even the large, densely populated river deltas of the Ganges, the Brahmaputra, or the Nile would be endangered. Some of the world's largest megacities lying close or next to the sea, such as Bangkok, Dhaka, Kolkata, Mumbai, New York, São Paulo, and Rio de Janeiro, would be at risk. Added to them are hundreds of large and small harbor towns and thousands of low-lying coastal areas. Many river deltas and coastal areas are among the world's most fertile agricultural regions. Even if they aren't flooded permanently, their harvests are threatened by floods, erosion, and groundwater salinity. Sewage disposal systems in many coastal cities could break down, with cholera, malaria, dengue fever, and the West Nile virus reappearing in large slums.

How high the sea level will actually rise by the end of the century and beyond is a hot topic at the moment. According to the cautious predictions in the 2007 IPCC report, we can expect an increase of eighteen to fifty-nine centimeters, depending on which SRES scenario is used as the base of reference. But the report potentially underestimates two important factors—ice melt in Greenland and the Antarctic region. More recent measurements indicate that the contribution of their melting ice sheets to rising sea levels will probably be much higher than surmised so far.

The sea level rose by roughly 1.5 to 2 centimeters per decade in the period from 1900 to 2000. Since 1993, with satellites providing far more accurate measurements, the seas have been rising by 3.4 centimeters per decade. This rate is likely to accelerate even more in decades to come if developments continue like those observed for several years by researchers at the Potsdam Institute for Climate Impact Research. They noticed that ice masses on the fringes of the Greenland and West Antarctic ice sheets, as well as alpine glaciers, are melting much more rapidly than most studies formerly predicted. If this trend continues, sea-level rise could well exceed 60 centimeters by the end of the century, and reach as much as 180 centimeters in a worst-case scenario.

Three main drivers are behind this rise in sea level—the thermal expansion of seawater, the melting of ice caps on Greenland and Antarctica, and receding alpine glaciers in non-polar regions. Researchers estimate that roughly 60 percent is caused by the thermal expansion of seawater, and the remaining 40 percent by melting ice caps and receding glaciers.

In addition to its impact on sea level, ice melt will affect living conditions for people, animals, and plants in a variety of ways. It will influence future climate development because the disappearance of ice-covered areas, now occupying roughly 10 percent of Earth's surface, will reduce albedo, the amount of solar energy reflected back to space. As ice-free water and snow-free land surfaces absorb more sunlight, ground level temperatures will rise, further accelerating global warming.

One of the most surprising findings in recent years is that Arctic sea ice is melting much more rapidly than expected. Between 1979 (when satellites were first used for collecting data) and 2000, summer sea ice areas occupied an average of more than 6.7 million square kilometers. By the summer of 2005, this figure had dwindled to 5.5 million and in the record year of 2007, it had gone down to only 4.1 million square kilometers. At the same time, sea ice thickness also diminished by at least 8 to 15 percent, with some data indicating reductions of up to 40 percent. If this trend continues, the Arctic will be free of ice during the summer months within just a few decades.

The Greenland ice sheet also seems to be melting faster than predicted. This body of inland ice is over three kilometers thick and rests on a layer of bedrock. But it is not static—under the weight of the snow that accumulates every year, ice constantly flows toward the sheet's margins. During the summer, the ice slowly melts or breaks off into the sea, forming icebergs, a process referred to as calving. Normally, the glacial system is balanced, with the annual amount of precipitation roughly making up for the amount of ice lost at the sheet's fringes in summer. But in a warming climate, ice melt and calving increase, causing the net volume of ice to shrink.

Glacial melting processes are very difficult to predict, partly because researchers were only recently able to start using satellites to obtain accurate data on the surfaces and volumes of the ice sheet's glaciers, and in particular the size of the melt zones. Computer models used by climate researchers to simulate these melting processes are therefore relatively inaccurate. More recent model calculations show that the mass of the Greenland ice sheet has been shrinking by 100 to 300 billion tons every year since the end of the 1970s, and that this rate of loss has accelerated since the 1990s. One sign of this is the expansion of the summer melt zones, which have increased by more than one-third over the last twenty years. In the summer of 2010 a new record was set when well over 40 percent of the total ice sheet was under melting conditions.

The situation in Antarctica is completely different. In winter, its ice-covered areas, including the ice shelves that jut out into the ocean, cover a surface of up to 30 million square kilometers—an area larger than Europe. It is very unlikely that Antarctica's inland ice sheet, with elevations of up to 5,000 meters above sea level, will melt; average temperatures are so far below the freezing point that even marked global warming would not cause the ice to melt. Because precipitation in Antarctica is expected to increase, the IPCC even predicted slight growth in the volume of the continent's ice masses in its 2007 assessment report.

Melting in Antarctica only occurs in ice shelf areas, where glaciers make contact with seawater. Massive ice shelf disruption has been observed repeatedly in these regions, especially in West Antarctica, where temperatures have risen by about 2.5 degrees Celsius (4.5°F) in the past fifty years. Nine major ice shelves have already disappeared, and the number of shelf-ice ruptures is evidently increasing. In February 2002, the Larsen B ice shelf, covering an area half the size of Germany, collapsed and broke into many larger and smaller icebergs. Climate scientists at the British Antarctic Survey are currently expecting to see a similar event at the Wilkins ice shelf. These breakups may cause glaciers feeding the ice shelf to accelerate. At ever shorter intervals, large sections of glaciers topple from the mainland into the sea. If this development speeds up in coming decades, it will contribute to a much higher rise in sea level than expected.

The retreat of alpine glaciers is well documented. It began as early as the nineteenth century, but has accelerated greatly in past decades. Glaciers in the European Alps have lost about one-third of their mass since the advent of industrialization, and if projections in the IPCC report prove to be accurate, glaciers in the Northern Hemisphere will shrink by at least another 60 percent on average by 2050.

Like in the Alps, most alpine glaciers in other parts of the world, in the Rockies, the Andes, the Himalaya and Pamir mountain ranges, and in New Zealand, have been retreating noticeably since at least the 1970s. If scientists' projections are correct, nearly all small glaciers in the world will have disappeared by the end of the century and total glacial mass will have decreased by about 75 percent. Many glaciologists believe that if the global temperature rises significantly more than 2 degrees (3.6°F) all alpine glaciers in the world will disappear within a few hundred years.

Data collected by the World Glacier Monitoring Service, a research center in the Department of Geography at the University of Zurich, show the speed at which alpine glaciers are currently retreating. An average of values measured over several years indicates that the world's glacial mass is currently melting nearly twice as rapidly as it did from 1980 to 1999.

According to the OECD, the number of people exposed to coastal flooding in large port cities could grow from 40 million to 150 million by the 2070s. This forecast is based on the assumption of a sea-level rise by 50 centimeters and increased storminess, both likely consequences of increased global warming and continued population growth and urbanization.

Both ocean warming and acidification are consequences of the CO_2 emitted into the atmosphere through human activity. These trends, in addition to marine pollution and overfishing, will lead to a further loss of marine biodiversity.

Glacier retreat alters countless river landscapes throughout the world, and to a greater extent the living conditions of people on river banks and in deltas. Some two billion people depend on rivers fed from the Himalayan glaciers—Asia's greatest rivers—the Yellow River, the Yangtze, the Ganges, and Brahmaputra, as well as the Mekong, the Salween, and the Indus.

The melting of inland glaciers brings about major changes that vary markedly over time. In the foreseeable future, river water levels will substantially rise, and retreating glaciers will leave behind moraines and glacial lakes that can suddenly drain and cause flash floods accompanied by scree avalanches and mudslides. Much later, probably not for a few hundred years, the overall amount of runoff from glacier melt will decline and its seasonal distribution will change. Rivers will swell very quickly during the spring snow melt, reaching abnormally high water levels, only to be followed by extremely low water levels during the summer months.

Both patterns may have enormous ramifications for large parts of eastern and southeastern Asia, for China, India, Pakistan, Laos, Cambodia, Burma, Vietnam, and Thailand, where large rivers supply entire regions with water and energy through extensive irrigation systems and dams. Many millions of subsistence farmers and fishermen who live along these rivers and whose livelihoods depend almost completely on yields from their small fields and river fishing grounds would be the most badly affected.

Another consequence of global warming is the thawing of permafrost in polar and alpine regions. On the basis of model calculations, the IPCC report estimates that roughly one-third of present-day permafrost in the Northern Hemisphere could thaw by 2080. There is already evidence of this in some high mountain valleys, where slopes have become unstable, causing landslides and rockfalls. The effect is also visible in northern Russia where large areas cave in because melting groundwater softens the ground, letting houses and streets sink into mud.

Damage caused by permafrost is still relatively minor, and the planning and construction of comprehensive protection measures in Austrian and Swiss mountain valleys started long ago. But the global potential for damage is enormous. Permafrost underlies 20 to 25 percent of Earth's land area, including 99 percent of Greenland, 80 percent of Alaska, 50 percent of Russia, 40 to 50 percent of Canada, and 20 percent of China.

There is a great deal of controversy among scientists regarding the degree of carbon dioxide and methane gas that would be released from frozen biomass when permafrost thaws. The total amount of carbon stored in permafrost is roughly 1,000 gigatons, and the total amount of methane is about 500 gigatons. If, in contradiction to nearly all scientific forecasts, a major share of these greenhouse gases were released, global warming would take on enormous dimensions.

But even without such apocalyptic visions, the prospect for coming decades until the end of this century is considerably more dramatic than the IPCC report's cautious and very general forecasts would lead us to believe, unless we succeed in adopting effective measures to reduce greenhouse gases very soon. But even stringent and swift measures will not save many millions of people, mainly in developing and emerging countries, from the severe consequences of global warming that are already inevitable. At the Cancún climate summit, Antonio Lima, ambassador of Cape Verde to the UN and vice-chairman of AOSIS, the Association of Small Island States, said, "We are going to be the first human species endangered in the twenty-first century." He appealed to delegates, saying, "We do not want to be sacrificed. We want to survive and to survive we need solidarity from those who can do something about the weather."

Glaciers calve when they flow into the sea, pushing ice toward the water until it breaks off. Perito Moreno Glacier in Patagonia, Argentina. *Steve Allen/Science Photo Library/Keystone*

Larsen B, an ice shelf in Antarctica more than 200 meters thick and 3,000 square kilometers in size, collapsed and broke away from the continent in March 2002. *NASA/Corbis*

The speed at which a glacier melts strongly depends on local conditions. Sogn og Fjordane county in southwestern Norway. *Bjørn Svensson/Science Photo Library/Keystone*

Melt water has carved a canyon fifty meters deep into a glacier in Greenland. Deposits of soot, volcano ash, and dust absorb solar radiation, accelerating the melting process. *James Balog/Aurora Photos*

Global warming is particularly visible in the Arctic, where permafrost prevents melt water from seeping into the ground, causing flat lakes to form. Hudson Bay, Canada. *Steve Jurvetson/Flickr*

The Zinal Glacier in Val d'Anniviers has lost most of its ice — just one example of how fast some glaciers are melting. Switzerland, 2009. *Françoise Funk-Salami/Keystone*

Maasai cattle wander across Lake Amboseli during the dry season in Kenya. Whether there will be enough groundwater for these herds hinges on two short rainy periods in spring and autumn. *Yann Arthus-Bertrand / Altitude*

Half of Burkina Faso is in the Sahel region, threatened by the advance of the Sahara; gardens and fields have been fenced to keep out livestock. *Georg Gerster/Keystone*

La Niña conditions have repeatedly caused droughts in southern Asia. A farmer works the soil in a dried-up rice paddy. Chongqing, China, 2009. *Stringer/Reuters*

Years of drought in southeastern Australia have made survival difficult for cattle breeders—a dwindling water hole near Nyngan, New South Wales, Australia. *Photo Library/Keystone*

The industrial cultivation of fruits and vegetables leads to ever greater water stress in dry parts of Spain and the eastern Mediterranean. Andalusia, Spain. *Yann Arthus-Bertrand/Altitude*

Only 46 percent of the Earth's coral reefs are still healthy today. Scientists suspect that global warming and ocean acidification will cause that percentage to rapidly decrease. Indonesia. *Dita Alangkara / AP / Keystone*

Rising sea levels threaten low-lying atolls; beaches and palm groves are washed away and salty seawater contaminates freshwater reserves. Majuro Atoll, Marshall Islands. *Chris Steele-Perkins/Magnum Photos*

A ROADMAP FOR THE FUT

REALITY POSSIBI

Martin Läubli

URE

AND

ITY

Since the 2010 United Nations Climate Change Conference in Cancún, Mexico, a simple number reflects official climate policy—2 degrees Celsius (3.6°F). This threshold represents the welfare of our planet. Climate scientists anticipate irreversible ecological damage if Earth's average temperature rises more than 2 degrees above preindustrial levels.

The two-degree target has far-reaching implications. Scientists at the Potsdam Institute for Climate Impact Research have estimated that this means no more than a fourth of today's commercially exploitable coal, oil, and gas reserves can be utilized by 2050. Greenhouse gas emissions, in particular the carbon dioxide (CO_2) released from heating systems, automobiles, factories, and fossil-fuel power plants, have been the primary cause of the marked warming trend of the last thirty years. Climate change challenges the very foundation of our wealth—the global energy system.

Today, politicians, economists, and scientists agree that the future belongs to energy sources that do not release any greenhouse gases into the atmosphere, and to technologies that consume as little energy as possible. The energy system of the future will encompass a diversified portfolio of energy sources such as hydropower, wind and solar power, and biomass. Despite the nuclear disaster at the Fukushima power plant following the earthquake and tsunami in Japan early in March 2011, nuclear power, controversial as it is, will most likely remain an option throughout the world. Geothermal energy supplies us with heat from deep below the ground; rooftop solar panels provide us with hot water. Plant oils and agricultural waste can be used to produce biofuel.

These technologies are already available and demand for them has exploded in recent years, especially for wind and

solar power. Yet energy experts still doubt whether the measures required to meet the two-degree target can be put into place fast enough. Energy production has evolved into a gigantic factory of CO_2 emissions, and grown nearly thirtyfold in the last hundred years. Almost two-thirds of the greenhouse gas emissions generated by human activity around the world come from the combustion of fossil fuels. The rampant waste of energy in industrialized countries in recent decades is mainly to blame. According to the Global Carbon Project of the Australian Research Institute CSIRO, CO_2 emissions have risen by 3.5 percent every year since 2000, which is nearly four times higher than the average yearly increase in the 1990s.

This development will become more and more extreme in the future unless we immediately take rigorous steps. The United Nations estimates that in about forty years more than 9 billion people will populate the planet; the global population today has reached 7 billion. Experts expect the largest population growth to occur in China, India, and Southeast Asia, and the fastest growth to be in Africa. Everyone will need access to affordable energy in the future to successfully combat poverty and improve public health. As it is, 1.5 billion people now live without electricity, and approximately 3 billion people cook and heat their homes with dung, wood, or kerosene, spending most of their day gathering the fuel they need.

The rate of economic growth in large emerging and developing countries will also play an important role. Experts at the International Energy Agency (IEA) believe that average growth in these countries will be double that of industrialized countries in twenty-five years. In 2008, the gross domestic product of Brazil, Russia, India, China, and South Africa equaled 25 percent of the gross global product. Their energy balance reflects this development. These countries currently consume a good third of the global energy supply and produce about 35 percent of the world's CO_2 emissions. Emerging and developing countries already release more CO_2 than industrialized countries do. Calculated on a per capita basis however, the affluent West still produces many times more—and this will not change any time soon.

The IEA predicts that unless effective political and economic incentives are created to make the switch to efficient and clean technologies, worldwide energy consumption will rise by nearly 50 percent by 2035. Consumption is expected to increase by 80 percent in emerging and developing countries, and by nearly 15 percent in industrialized countries.

Against this background, is it possible to revolutionize an energy system, which has worked reliably for nearly a century, within just forty years? Government and industry face a dilemma. On one hand, the dangers of climate change demand rapid, effective measures to introduce CO_2-free and environmentally sustainable energy. On the other hand, the hunger for electricity and heating due to population expansion and economic growth is so great that, for now, high-growth countries such as Brazil, Russia, India, and China are counting on conventional and proven fossil-fueled and nuclear power production. The opposite is true in industrialized countries; they do not expect energy demand to increase much in the future and therefore they focus more on developing sustainable energy systems.

Consequently, prognoses regarding the power mix of the future vary widely. The World Energy Council and the IEA both say that in spite of a worldwide boom in renewables, fossil energy will continue to play a dominant role in coming decades. They point out that the investment needed for climate-friendly technologies is much higher than that needed for constructing or retrofitting fossil-fuel power plants →p.401. At the 2010 World Energy Congress, the energy industry therefore advocated investing in controversial CCS (carbon capture and storage) technology as the only way to effectively reduce CO_2 emissions. This process involves filtering CO_2 out of fossil power plant emissions and storing it underground. However, its large-scale commercial use is not expected to begin for another twenty years, if ever →p.303.

The International Renewable Energy Agency (IRENA) holds a more optimistic view of the future, believing that renewable energies will contribute to 50 percent of the global mix by 2050. Today their share is roughly 18 percent. Scenarios developed by the European Renewable Energy Council and the environmental organization Greenpeace are even more optimistic. They believe that a share of up to 80 percent is realistic and predict that 95 percent of global electricity will be generated from renewable energy sources even though demand for electricity will experience a stronger rise in the future than other forms of energy. They expect the energy revolution to be affordable and to create millions of new jobs. The main obstacle in their opinion is the lack of political will.

But for now, energy experts assume that countries will generally attempt first to exploit domestic reserves to ease their dependency on energy imports. China and India will rely on coal, and Canada on bitumen extracted from oil sands in a very complex process that lays vast areas of land to waste. Countries like Japan, South Korea, and France, which do not have any resources, will rely on nuclear energy to generate electricity, and the disaster in Japan means that the nuclear industry will have to try even harder to persuade the public that its power plants are safe. On the other hand, the European Union is planning the long-term development of power generated exclusively by renewable energy sources. This plan foresees a grid that will combine large-scale centralized power plants such as wind farms, hydropower plants, and solar thermal facilities with decentralized energy sources including photovoltaic modules on rooftops, biogas plants on farms, and small hydroelectric power plants in municipalities.

The issue of mobility always tends to be ignored however in any debate on restructuring our energy system →p.337. The question is not only whether cars in the future will be powered by fuel-efficient combustion engines or electric drives, but also what kind of mobility is compatible with a sustainable future. Thousands of new communities will spring up around the world in coming decades, and megacities and economic areas will continue to grow. The global fleet of vehicles could triple by 2050, with most new cars and trucks on the road in developing countries. City and regional planning and the expansion of public transportation networks will play a key role →p.357.

The restructuring of our global energy system is the most important project of the century. Climate researchers and environmental economists estimate that it would be most cost-effective if CO_2 emissions peaked in ten years at the latest, and then decreased by 50 to 80 percent by 2050 →p.401. This calls for additional, more immediate solutions. Energy efficiency is the area where we can expect to see the most progress in the short term. This also seems to be the only common goal at the moment among various global energy interests. United Nations data indicate that improved efficiency in the production of energy and in industrial operations could lower by more than half those additional energy needs expected for the next twenty years. If relevant standards were enforced around the world, then we could reduce greenhouse gas emissions by roughly one-sixth.

Climate change, however, raises questions that even green politicians no longer ask. What are the limits to growth? Scientists in the 1970s made headlines when they warned against exploiting the planet. Even though computer models in those days provided only rough figures, nothing has changed as far as the validity of their message is concerned. Growth has always undermined measures to mitigate climate change. Progress in energy efficiency is frequently offset by even larger economic growth. Economic development will force emerging and developing countries to rely on cheap fossil fuels—unless the Western world provides poorer countries with financial and technological support, enabling them to invest in a climate-friendly energy system. Moreover, technologies that are allegedly sustainable can also devour resources. Biofuels made from corn and soybeans, for instance, are not an alternative to fossil fuels because they lay claim to huge areas of agricultural lands needed for food production →p. 392. Manufacturing solar cells and batteries for electric cars makes no sense if it means using rare earth metals that are of limited availability.

Engineers are counting on the innovative human spirit in the battle to mitigate climate change. It is however doubtful whether technology alone will turn things around. It will be nearly impossible to avert critical levels of global warming without cutting back on our consumption of energy. We need financial incentives to achieve this, either in the form of government subsidies or higher energy prices so that homeowners will be motivated to insulate their houses and invest in climate-friendly heating systems. It is also important that consumers get into the habit of taking environmental standards into consideration when they are shopping. Another decisive factor will be how often people drive and how frequently they refrain from flying →p. 411.

Mitigating climate change is the greatest challenge of the century. Many climate scientists are skeptical that we will be able to avert global warming of more than 2 degrees

Celsius (3.6°F). Poor countries in Africa and Asia already suffer from floods, storms, and droughts on a regular basis, natural disasters that could become more frequent in the future as an outcome of climate change. These countries experience economic and social setbacks with every disaster that strikes. Those who have the means adapt to climate change by building dams and protective structures or by cultivating crops that need little water →p. 372.

If greenhouse gas emissions are not decisively curbed in coming decades, then the risk grows that adaptation will gradually become ineffective and the resulting damage will be overwhelming. We are at a crossroads. Ultimately, the restructuring of our energy system depends on political decisions and economic interests. Many power plants in Europe and the United States are nearly at the end of their operating lives, whether they are coal-fired, gas-fired, or nuclear. Plans for the coming ten years will determine whether we initiate changes or continue to rely on CO_2-intensive technologies, which, as we know, contribute to global warming.

 Unbounded demand for energy — a sea of lights in Las Vegas. *John D. Norman/Corbis*

Artificial winter in desert heat — more than 6,000 tons of artificial snow create this 400-meter ski run at an indoor ski resort in Dubai. *Ahmed Jadallah/Reuters*

Construction site on the outskirts of Jammu city in the state of Jammu and Kashmir, India. Population growth and increasing prosperity will spur huge infrastructure expansion over the coming decades. *Jaipal Sing/Keystone*

An opportunity for better, climate-friendly building — high-rise apartment buildings are torn down in Shenzhen in 2005 to make room for a new city development plan. *Tsa Ka Yin/Keystone*

Billboard promoting a better life in Dubai, the largest city in the United Arab Emirates. *Jeremy Horner/Corbis*

Eager interest in greener and more comfortable living—public presentation of a residential development project in Shanghai. *Imagechina/Keystone*

 Higher income fosters the desire for mobility — an automobile plant in South Africa. *Charles O'Rear/Corbis*

Higher income also means growing consumer appetites, especially in countries with booming economies—a shoe factory in China. *Hiroji Kubota/Magnum Photos*

Globalization has accelerated the release of greenhouse gas emissions—a container terminal in the port of Hamburg, Germany. *Roland Magunia/Keystone*

A tollgate in France—most goods continue to be shipped by truck. *Stephane Ruet/Corbis*

Consumer products imported by air freight have a particularly negative CO_2 balance. Cargo terminal at Chek Lap Kok Airport in Hong Kong. *Vincent Yu/Keystone*

Growing consumerism in developing countries causes energy consumption to soar toward Western levels. Lake Palace, a shopping mall outside of Tunis, Tunisia. *Harry Gruyaert/Magnum Photos*

Year-round summer comes at the expense of lavish power consumption — one of the world's largest indoor water parks, Miyazaki City, Japan. *Hiroji Kubota/Magnum Photos*

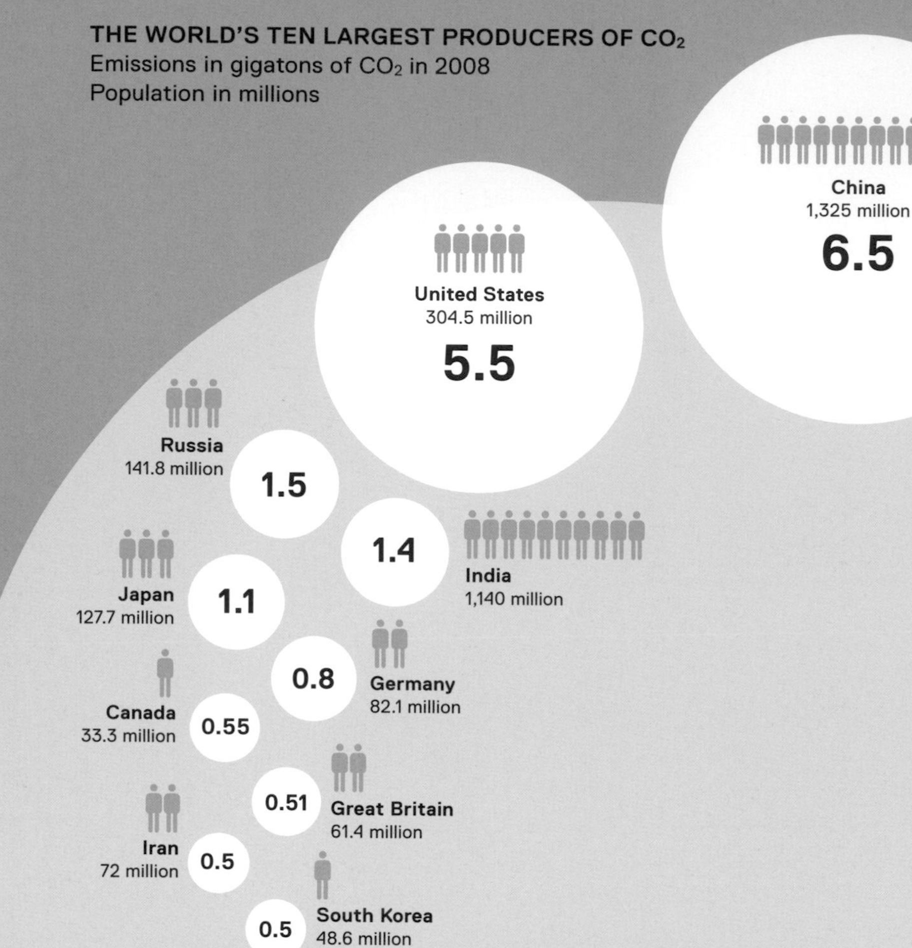

Source: IEA CO_2 Highlights, 2010; United Nations Statistics Division, 2010

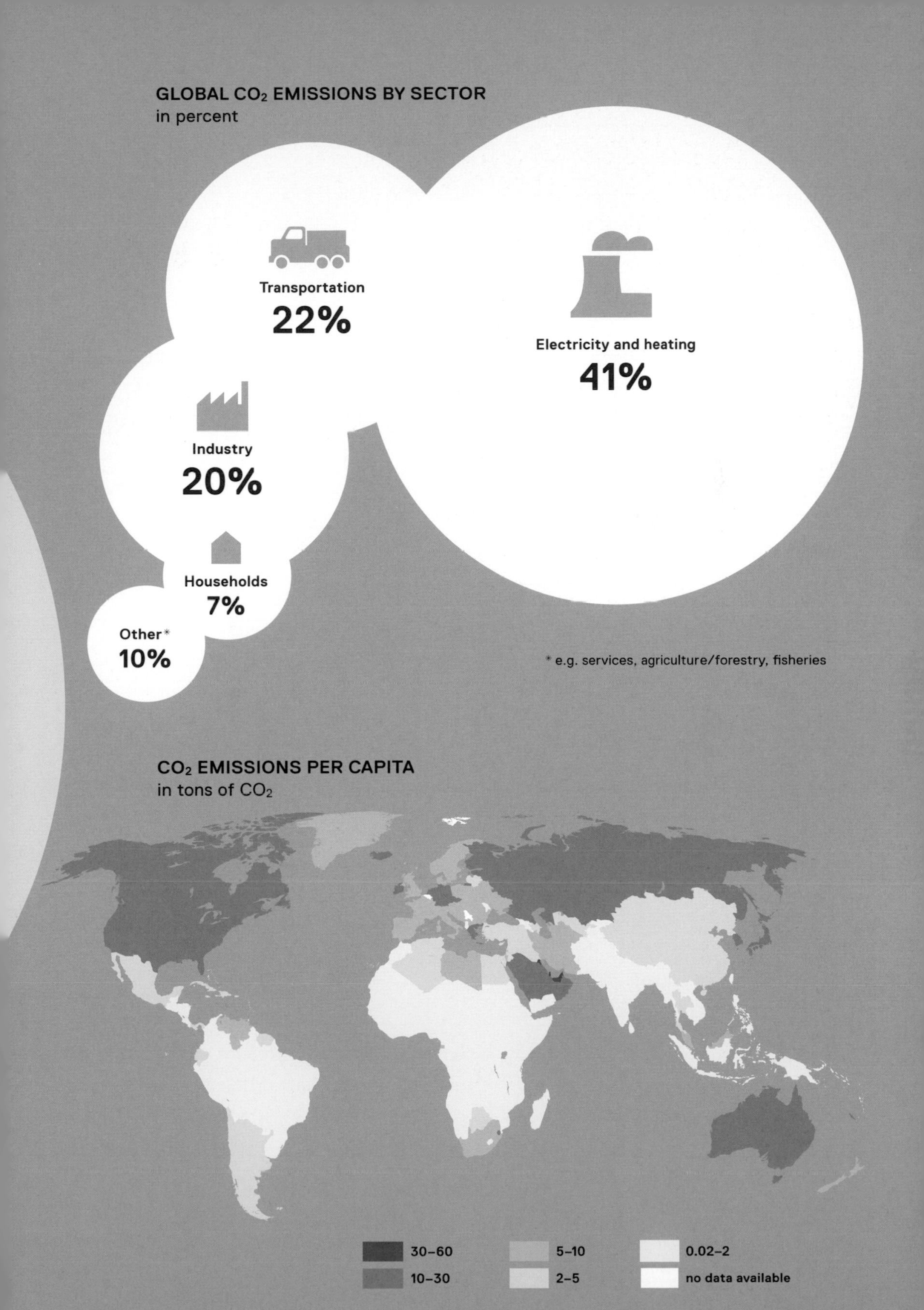
GLOBAL CO_2 EMISSIONS BY SECTOR
in percent
Transportation
22%
Electricity and heating
41%
Industry
20%
Households
7%
Other*
10%
* e.g. services, agriculture/forestry, fisheries
CO_2 EMISSIONS PER CAPITA
in tons of CO_2
30–60
10–30
5–10
2–5
0.02–2
no data available

DOWN TO THE LAST DROP OF OIL?

The substance invariably evaporated slowly, burned brightly, and was not all too expensive. In the middle of the nineteenth century, American industry discovered kerosene—distilled from oil—as an excellent lighting material. When Edwin Drake drilled the first oil well in 1859 in Pennsylvania, the breakthrough for oil, at least for lighting, was only a question of time. The most popular oil used in lamps in America at that time was camphene, a mixture of alcohol and turpentine. Methyl alcohol was distilled from wood, and turpentine from pine resin. Lamp oil became more expensive between 1862 and 1864 when the American government put a tax on alcohol to replenish the war chest for the Civil War.

But some time passed before oil was established as a fuel. Legendary American automaker, Henry Ford, was convinced that alcohol produced from fruit, grass, and vegetables would be the fuel of the future. He said this as late as 1925 to *New York Times* reporters; Ford was concerned about ailing American agriculture.

When the first cars with internal combustion engines came on the market, scientists explained in hundreds of papers why alcohol was preferable to gasoline, aware that alcohol could be produced simply from plant waste, and that engines knock less with alcohol and don't leave behind fumes. These were persuasive arguments, especially for France and Germany, since they had scant oil reserves. At the turn of the twentieth century, scientists and governments in Europe were also worried about how long the oil resources of Russia and the United States would last.

But in spite of forceful technical arguments, the oil industry lobby prevailed in the end. Oil production was cheaper. The discovery of lead tetraethyl was significant—it proved to be an outstanding antiknock agent and was mixed with gasoline as an additive, even though experts pointed out the health risk of leaded gasoline. As we know, only lead-free gasoline is in use today.

Cheap and easy to transport, oil gradually replaced coal during the Industrial Revolution. John D. Rockefeller founded the international company Standard Oil in 1870, becoming the "father" of the oil industry and earning a billion dollars within fifteen years. The age of oil began. In the United States, oil groups like Texaco and Gulf Oil were established. Shell discovered oil in Indonesia, and British Petroleum in Iran. The world's second largest oil field was discovered in Kuwait in 1938, and the largest in 1948 in Saudi Arabia. With the discovery of oil in the Middle East, global production exploded after 1950, rising by 7 percent per year for the next twenty years.

The Organization of Petroleum Exporting Countries (OPEC) was founded in 1960 in Baghdad. Its members today are Iraq, Kuwait, Saudi Arabia, Qatar, Libya, United Arab Emirates, Algeria, Nigeria, Angola, Ecuador, Venezuela, and Iran. OPEC states produce about 40 percent of the world's supply of crude oil and possess more than three-quarters of global oil reserves. The International Energy Agency (IEA) estimates that OPEC's share in production will increase to more than 50 percent by 2035.

OPEC's influence became visible during the first oil crisis in 1973. The oil-producing states deliberately cut back production to protest against Western support of Israel in the Arab-Israeli Yom Kippur War, and the price of oil rose by 70 percent. In 1979 and 1980, production losses and a global sense of uncertainty about future energy security after the revolution in Iran and the first Gulf War between Iraq and Iran caused another price increase. Uncertainty today is generated by democratization processes in Arab states, which began in Tunisia and Egypt and spread across the entire Arab world, and which now involve oil-producing countries such as Libya. So far, this development has had only a marginal effect on the price of oil.

The oil crises clearly revealed the vulnerability caused by oil dependency. In the early 1970s, the Club of Rome published *The Limits to Growth,* whose warnings against the wasteful consumption of natural resources received worldwide attention. In 1972, at the initiative of the Soviet Union and the United States, scientists from around the world founded the International Institute for Applied Systems Analysis (IIASA) in Laxenburg, Austria, to comprehensively analyze the present and future of global energy supply. The summary of their report, written more than thirty years ago, is more relevant than ever. Researchers believed it was possible to set up a practically inexhaustible energy system in which electricity and hydrogen would be the energy carriers of the future. Solar installations and nuclear power plants would supply the power.

The all-important question has been posed for decades: How long will oil last? For scientists, the crucial factor is not the volume of reserves in deposits, but rather the point of maximum production, the peak. When an oil field reaches its peak, the pressure of deposits has decreased so far that the extraction rate continually decreases. The point when this becomes the case for all global deposits is what is referred to as "peak oil." Geologists aren't able to determine this point in time with precision because different factors play a role in forecasting maximum global oil production, such as the volume of extractable oil available, including new finds, and feasible daily production in the future. Scientists often look at past production trends. Geophysicist M. King Hubbert worked with data on U.S. oil production in the fifties and concluded that maximum production in the United States would occur in the early seventies—and he was right. The accuracy of prognoses for global peak oil is also limited by the fact that extraction rates in states like Saudi Arabia, holding more than a quarter of conventional—i.e., economically viable—oil reserves, are difficult to estimate. OPEC policy is another unpredictable factor. There is also uncertainty about continuing access to important oil fields in Iran and Iraq in the future. Not least, consumer behavior is crucial. The international consulting firm IHS CERA assumes that demand for oil will not rise in coming decades in North America and Europe, but that consumption in developing and emerging countries will continue to rise because of increased mobility. Overall, oil consumption is still expected to go up worldwide.

The largest oil fields were discovered more than fifty years ago. The number of new oil fields discovered since the sixties has diminished around the world. In the United States, excluding Alaska, the maximum number of new discoveries was reached in 1932; forty years later peak was reached in oil production. In 1974, the number of newly discovered oil fields in Great Britain fell off, and twenty-five years later oil production diminished after peak was reached in the North Sea in 2000. Global oil production on land has been decreasing since the nineties, and only the extraction of offshore oil has stopped the downward trend.

Whenever it comes, the global maximum production point will ultimately occur. Since the beginning of the nineties, we have been

Daily consumption of crude oil in 1970 was 9.6 million barrels. By 2010 it reached 85 million barrels. World production rates are stagnating. About half of the conventional reserves have already been used up, and current oil consumption rates cannot be sustained much longer.

consuming more oil than we are finding in new deposits. The International Energy Agency (IEA), whose estimates are conservative, officially admitted for the first time in 2010 that the world's conventional oil production had peaked. But if the price of oil continues to remain as high as it is today or goes even higher, unconventional oil fields that are difficult to access, making their exploitation too costly, will become attractive. Canada and Venezuela are already counting on extracting oil from tar sands in an expensive but still affordable process, which, however, massively damages the environment. Another source is oil shale in sedimentary rocks, which holds kerogen, an organic material which must first be intensely heated to a high temperature for oil to be released. More than 50 percent of the new oil and natural gas reserves in the world have been discovered in the Gulf of Mexico and off the east coast of South America, and deep sea extraction has tripled in the last ten years. Experts at Germany's Federal Institute for Geosciences and Natural Resources believe this trend will continue—as long as high safety standards are applied. But other experts doubt that this oil, difficult to access, can meet future demand. The 2010 Deepwater Horizon oil spill in the Gulf of Mexico shows that such a disaster can take even an oil giant like BP to the brink of ruin.

The German section of the Association for the Study of Peak Oil and Gas (ASPO) has observed that when less oil is available, individual oil producers have no choice but to merge to keep up the appearance of growth. Investment firm Goldman Sachs noted as early as 1999 that mergers of oil groups are the sign of a dying industry.

German economist Hans-Werner Sinn believes that oil corporations would do anything to get the last drop out of the ground and sell as much oil as possible. At the same time, increasing demand for renewable energies like wind and solar power could lead to dwindling interest in oil, making oil cheaper and more attractive to countries that could not afford expensive alternative energies. Sinn warns that less demand from "green" EU countries would effectively subsidize oil for the United States and allow Americans to drive even bigger cars than they do anyway, and China to further speed up its CO_2-intensive growth. He believes climate change can be mitigated only if the owners of fossil fuels can be persuaded to leave deposits underground.

Although this generally seems to make sense, it also seems a naive demand in the geopolitical context. Reality also speaks against Sinn's thesis. During the recession of 2008–9, for example, when oil consumption decreased in the European Union and the United States, China did not buy up all the cheaper "unused" oil to satisfy its enormous hunger for energy. Quite on the contrary, it increased domestic oil prices, limited the use of diesel, and restricted vehicle registrations. Furthermore, the oil market does not always operate tightly in line with the economic principles of supply and demand. OPEC can always prop up prices with the help of production quotas.

If we use IEA prognoses as a standard, then more coal than oil will be consumed in the future. Coal can be liquefied to fuel—South Africa has been processing coal this way since 1955, although it is the only country in the world to do so. The World Coal Association assumes that coal reserves will probably last for another 120 years. American scientists at the Post Carbon Institute in California have warned however that an energy policy relying on a nearly unlimited supply of coal needs to be "urgently" reconsidered. Indeed, various new studies paint a more pessimistic picture. Some of these assume that global maximum production for coal is no longer decades away, but only years. Chinese scientists have looked at historical trends for coal production in China and have predicted peak coal for 2025. Their prognosis is based on figures for conventional reserves, which the government now officially estimates at 187 billion tons. But such estimates are disputed because production depends not only on geological reserves, but also on the quality of coal and the cost of supply. Even China with its enormous coal reserves periodically imports more coal than it exports. This can happen when the delivery of coal from Chinese interior provinces to buyers on the coast is more expensive than the import of coal from Indonesia or Australia.

Meanwhile, demand for coal continues to increase. China, India, Australia, Poland, and South Africa today obtain 70 to 90 percent of their power and heating from coal-fired plants. The International Energy Agency (IEA) forecasts that China's industry alone will burn 50 percent more coal by 2035 than it does today →p. 303. Experts believe this enormous demand will use up reserves much more rapidly than previously assumed. Because national data are often unreliable, estimates generally vary widely. When experts speak of reserves, they mean those seams of coal that geologists say can be mined in a technically and economically viable way. Figures for reserves could be higher if new mining technologies and higher prices for coal were to make more difficult extraction profitable.

But researchers at the Post Carbon Institute say that this sort of development hasn't been observed anywhere except in Indonesia and India. On the contrary, between 2003 and 2008, reserves decreased by one-third in Germany and South Africa, for instance. In the United States, geologists recently estimated that the country has enough coal for the next 240 years, conceding however that data are still imprecise. Statistics from the late nineties show that although the volume of coal extraction continued to increase, the quality decreased overall, and with that, the energy content of the coal.

Whether peak oil or peak coal—maximum production does not say anything about how long the reserves of a raw material will last. We would have to know how fast the production rate would drop after peak and how high demand would be. The production curve could also plateau in the event that more unconventional oil or coal could be extracted by introducing new, less expensive production techniques. The outcomes of various studies on oil production rates, whether from Shell, OPEC, or the IEA, diverge widely. Prognoses on conventional oil reserves range broadly—from twenty to forty years. The World Coal Association assumes that coal reserves will last for another 120 years or so.

Nevertheless, the decisive factor for energy provision of the future is not the reserves in deposits, but maximum production. The peak would have an effect on oil or coal prices because increased demand could no longer be met. The German section of ASPO is not alone in issuing a warning about peak oil. Even the German armed forces, the Bundeswehr, has drawn up a report on the subject and is concerned that maximum production could have major systemic consequences. Fuel as well as 95 percent of all industrially manufactured products (ranging from plastics and pharmaceuticals to dyes and textiles) now depend on the availability of oil. An increase in oil prices would affect all areas of the economy and society, including the automobile industry, the building trades, and tourism. The Bundeswehr study concludes that the end of cheap oil could have drastic consequences, leading to recession, increasing unemployment, and the collapse of the financial system.

International climate policy is trying to press ahead with the transition from fossil fuels to renewable energy sources by prescribing the reduction of greenhouse gas emissions. National energy supply plans also consider using gas-fired power plants as a bridging technology until a broader use of renewable energy sources becomes economically viable. Experts at the 2010 World Energy Congress in Montreal believed that natural gas would play an important role in future power production because gas-fired cogeneration plants, producing both heat and electricity, today achieve 60 percent efficiency. The United States, for instance, could approach medium-term climate targets alone by replacing obsolete coal-fired power plants with gas-fired plants, simply because burning gas produces considerably less CO_2. Furthermore, the United States is expected to extract shale gas more intensively in the future because advances in production technology in recent years have now made exploitation affordable. China and Russia are also pursuing the exploration of natural gas. This could lead to available global reserves of natural gas greatly expanding in the next few decades.

It would make sense to massively curtail subsidies on fossil energies in the next few years to speed up advancements with renewable energy sources and to mitigate climate change. The IEA says that global subsidies for oil, coal, and natural gas came to more than $550 billion in 2008 and has calculated that demand for fossil fuels without such support could go down by 6 percent and global CO_2 emissions by 7 percent. That would be even better than the goal set in the Kyoto Protocol.

The collateral damage of modern lifestyles — oil production destroys entire landscapes, here near Baku, Caspian Sea, Azerbaijan. *Ian Berry/Magnum Photos*

The fragmentation of landscapes is a secondary concern whenever oil is discovered—oil fields near Puesto Hernández in the Argentine province of Neuquén. *Yann Arthus-Bertrand/Altitude*

Ugly scars, profound damage—tar sands extraction at a mine in Fort McMurray, Alberta, Canada. *Orjan F. Ellingvag/Dagens Naringsliv/Corbis*

China relies heavily on coal for its electricity supply, making it the world's largest CO_2 emitter. Pingshuo Antaibao coal mine in Shuozhou, 2008. *Lo Mak/Corbis*

Brown coal open-pit mining near Jänschwalde, the Vattenfall coal-fired power plant near Cottbus—a symbol of our dependency on fossil fuels. *Fabrizio Bensch/Reuters*

No quick remedy – insufficiently insulated buildings will continue to consume enormous amounts of heating fuel for decades to come. A winter's day in Moscow. *Barry Lewis/Corbis*

Oil is everywhere — oil-based household products on sale at a market in Ecuador. *Peter Guttman/Corbis*

Short life cycles — plastic bottle recycling in Tallinn, Estonia. *Mika/Corbis*

Approaching Western standards — booth at the 106th China Import and Export Trade Fair in Guangzhou, China. *Lu Hanxin/Corbis*

Oil is an inexpensive raw material — plastic chairs in Zurich, Switzerland. *Alessandro Della Bella/Keystone*

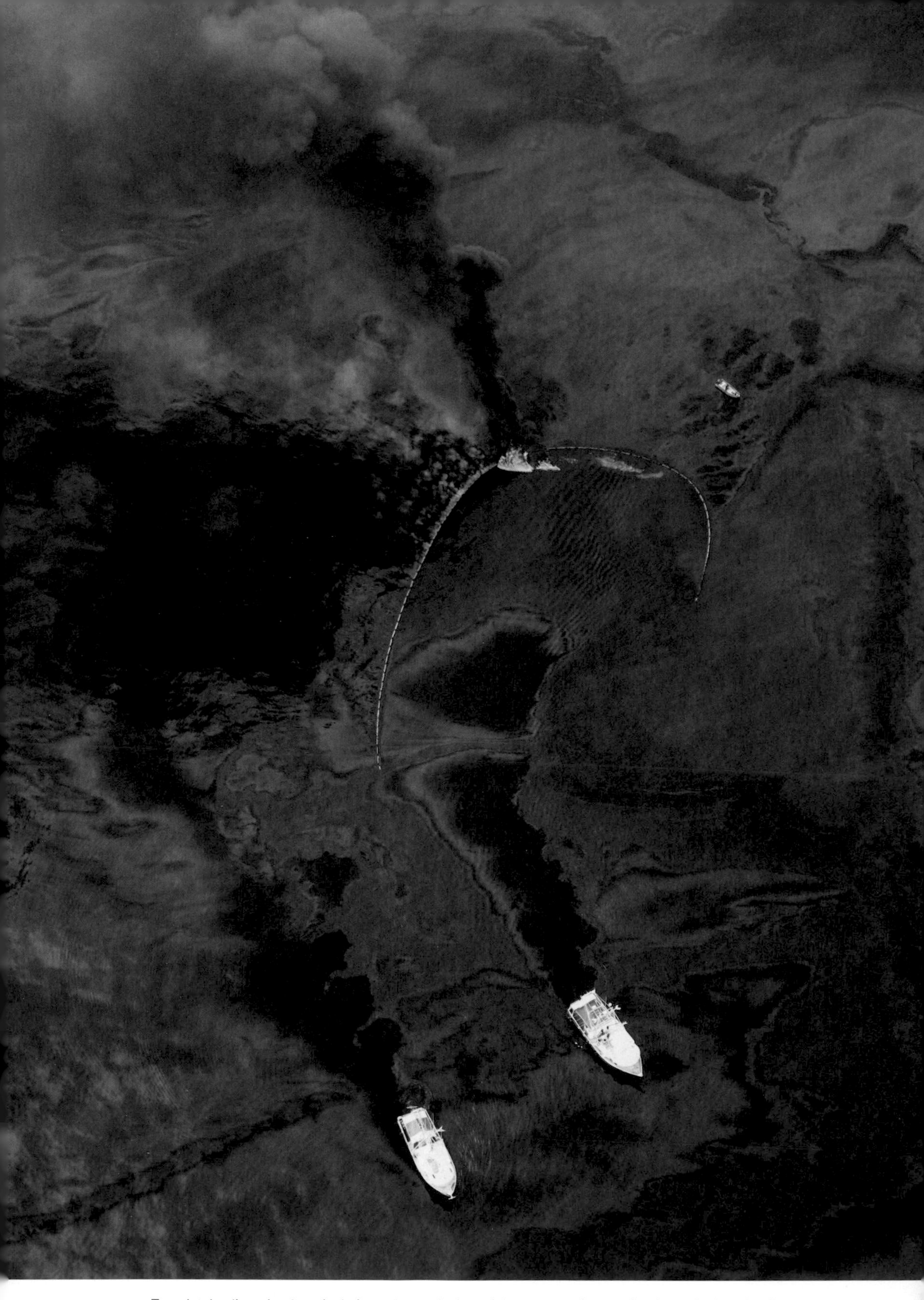

To maintain oil production, the industry increasingly exploits reserves deep under the seabed — oil spill in the Gulf of Mexico in 2010 following the explosion of the Deepwater Horizon drilling rig. *Eric Gay/Keystone*

THE HOPE FOR CLEAN COAL

Coal is more detrimental for the climate than any other source of energy. No other fossil fuel releases as much CO_2 into the atmosphere as brown coal and anthracite when they are burned to generate electricity. Capturing carbon dioxide from the emissions of coal-fired power stations and storing it deep underground or in the ocean has been proposed as a way to solve this problem. Called CCS (carbon capture and storage), this process however still faces enormous technical and economic difficulties as well as public opposition.

Klaus Lanz

Coal has one great advantage—it is widely available around the globe. The world's coal reserves are much more abundant than reserves of oil and gas. Yet coal's negative impact on climate has made it less and less acceptable as a fuel and has put the coal industry under pressure to come up with an environmentally compatible solution for its future use. CCS technology has sparked a heated debate in Europe, especially in Germany, where anthracite and brown coal currently generate nearly half of the country's electricity supply and are responsible for roughly 80 percent of CO_2 emissions related to power. The leading coal advocates in Germany are Vattenfall, the Swedish energy company that mines and processes brown coal reserves in former East Germany with its subsidiary Mitteldeutsche Braunkohle, and RWE with Rheinbraun, its brown coal subsidiary. Outside of Europe, CCS is being pushed by China, Australia, and the United States, the countries with the largest coal reserves and the highest dependency on coal for power production. Coal is the number one fossil fuel around the globe, far ahead of all others when it comes to generating electricity.

Why shouldn't it be possible to filter CO_2 from power plant emissions the same way that nitrogen and sulfur oxides have been scrubbed ever since acid rain became an issue? The answer is simple—while emissions contain only traces of sulfur dioxide (SO_2) and nitrogen

In an attempt to improve the CO_2 balance of coal, Vattenfall tests the capture of carbon dioxide from flue gas. Vattenfall's Schwarze Pumpe brown coal power plant in Spremberg, Germany. *Andreas Muhs/Keystone*

oxides (NO_X), the main product of combustion is actually carbon dioxide (CO_2). Every ton of coal burned produces 3.7 tons of CO_2, but only a few kilograms of SO_2 and NO_X. A simple calculation illustrates the problem: a 1,000-megawatt coal-fired power plant equipped with CCS technology would burn about 9,000 tons of coal per day. This means it would produce 33,000 tons of CO_2, which would have to be captured, carried away in pipelines, and finally injected deep underground, day after day. In a year, this adds up to 12 million tons of CO_2. Extrapolated to include the total amount of coal burned in Germany today, the annual figure would be twenty-five times higher.

The amount of CO_2 contained in waste gas or flue gas from coal-fired power plants is at most 14 percent. In principle, it could be separated using various physical and chemical processes. However, these technologies share the problem that they are all extremely complex and energy-intensive, and make power generation significantly more expensive. A plant equipped with CCS needs to burn about 30 percent more coal to produce the same amount of electricity. Separating the main components of flue gas, carbon dioxide, and nitrogen consumes a lot of energy. This is not a matter of inefficient technology, but rather of some basic principles of physics. Large coal-fired power plants produce millions of cubic meters of flue gas per day, and constructing suitable facilities to process such huge gas volumes is much too costly to ever make retrofitting existing power plants economically viable.

CCS TECHNOLOGY REQUIRES NEW POWER PLANTS

That's why existing power plants are not part of the industry's plans for CCS. It is interested only in constructing new power plants equipped with new technologies such as the oxy-fuel process, which burns coal using pure oxygen instead of ambient air. About 80 percent of the flue gas from this process is CO_2 and the other 20 percent is water vapor, two gases that can be easily separated.

However, oxy-fuel combustion technology is still at a very early stage. One particular problem involves finding adequate combustion chamber materials, because burning coal with pure oxygen calls for the use of very durable, chemically inert equipment. Another problem yet to be solved is how to generate huge quantities of oxygen at affordable prices. If these obstacles could be overcome, commercial oxy-fuel power plants could go into operation starting around 2030.

Another type of technology is being tested in integrated gasification combined cycle (IGCC) plants. In this process, coal is gasified before being burned. The gasified coal, or syngas, reacts with water vapor under pressure and at high temperatures to generate a mixture of CO_2 and hydrogen. The CO_2 can be captured from the syngas while the hydrogen can be used as a carbon-neutral fuel to generate electricity.

This type of power plant is much more efficient, with a much higher yield of electricity than that from conventional combustion technologies. The Chinese coal industry in particular is counting on this process. However, IGCC is still at a very early stage of development. A research power plant in the Chinese city of Tianjin scheduled to go into operation in 2011 will show whether this process is at all suitable for industrial, large-scale use. Even if development continues at a rapid pace, however, it is likely that it will be another twenty years before IGCC power plants are ready to operate commercially.

HOW CAN WE GET RID OF CO_2?

The technical challenge of generating electricity from coal without emitting CO_2 is already huge. But dealing afterward with liquid carbon dioxide itself is even more demanding because enormous tonnages have to be safely stored deep underground for centuries to come.

The easiest solution would be to inject liquid CO_2 into depleted oil and gas fields, which have been well surveyed after decades of use. But even optimistic estimates by the International Energy Agency predict that the capacity of these fields would be saturated in twenty years at the most. Another complication is that the majority of former oil and gas fields in Europe are located offshore, where CO_2 storage is twice as expensive as on land because of the need for more pipelines, drilling, and platforms.

The coal industry is therefore considering depositing CO_2 in repositories 800 meters and more underground, in rock layers filled with concentrated salt water that lie under large parts of the Earth's continents. Researchers do not know much about the geological features of these deep saline aquifers. No one can tell today if and to what degree they are suitable for long-term storage. The Norwegian energy company Statoil is carrying out a pilot project in the North Sea, injecting carbon dioxide into a sandstone formation 800 to 1,000 meters below the seabed. Seismic measurements indicate that the CO_2 gradually spreads out in the depths. It is nearly impossible to verify whether it stays there permanently or whether part of it leaks back up into the atmosphere. Besides, the amount of CO_2 that Statoil is storing is not significant compared to the CO_2 tonnage produced by coal-fired power plants.

The U.S. Department of Energy estimates that 3.3 to 12.6 billion tons of carbon can be stored in deep, underground saline aquifers in the United States and Canada, but no practical experience has been gained with storing carbon dioxide in this way on the North American continent. Since the American power industry relies on coal to generate more than 50 percent of its electricity, there is intense political pressure to find a way to store captured CO_2. Australia is also looking actively for potential repositories for CO_2. As of 2011, the American oil corporation Chevron is planning to attempt depositing up to 3.4 million tons of CO_2 per year 2,500 meters under Barrow Island, located fifty kilometers northwest of the Australian continent.

Indeed, repositories have to be as close as possible to power plants because building pipelines to transport CO_2 to storage locations significantly increases the cost of power production. Maximum distances of 300 kilometers are still considered economically viable. This essentially means that most CO_2 must be injected into the ground on land-based sites. Authorities in Germany, for instance, have already identified 408 potential locations, about half of them under the mainland. Local residents are vehemently opposed

Coal reserves are abundant; however, their extraction can be justified only if coal-fired power plants are developed that don't emit CO_2. Open-pit coal mine near Ava, Illinois, USA. *Charles Rex Arbogast/AP/Keystone*

to this kind of storage as geological reports have concluded that CO_2 injected in saline aquifers could push up salt water from deeper layers and cause large-scale contamination of overlying groundwater. This is also the reason why water utilities in northern Germany, who abstract drinking water from those groundwater reservoirs, are fiercely resisting the deep underground storage of carbon dioxide.

THE FUTURE OF COAL

The commercial use of CCS will have to wait for a while. The Massachusetts Institute of Technology does not expect significant applications to be in place before 2030, and according to a special report by the IPCC, CCS systems are not likely to be in use until the second half of the century. For at least another twenty years, most of the electricity in countries that continue to rely primarily on coal will have to be generated in conventional power plants—CCS will not mitigate CO_2 emissions until 2030 at the earliest. UNDP, the United Nations Development Programme, concludes that at the expected rate of deployment, CCS "will arrive on the battlefield far too late to help the world avoid dangerous climate change."

Today's political decisions will determine the energy mix of the next thirty to forty years. If we want to deal with climate change effectively, we must concentrate first and foremost on intelligent CO_2-free technologies that are already available. Investing in mature technologies for renewable energy today would diminish CO_2 emissions much more quickly than waiting for CCS.

The powerful coal lobby has succeeded in securing comprehensive government support in many countries for its CCS research and development. At the same time, financial support for the development of renewable energy is on the wane. Australia runs three research facilities for fossil fuels, one of which is dedicated entirely to CCS, but it does not have a single research center focusing on renewable energy. And in the United States, in 2010 alone the Department of Energy made more than $4 billion available for CCS programs.

Electricity producers in the West are considering shifting coal-burning operations to other countries in response to public opposition directed at the underground storage of carbon dioxide. E.on, a German group and the world's largest privately owned power corporation, is considering investing in coal-fired power plants in Russia. At the 2010 climate conference in Cancún, the capture and storage of CO_2 was included in the Clean Development Mechanism (CDM) of the Kyoto Protocol. In the future, this could actually create a situation in which rich signatories of the Kyoto Protocol invest in CCS coal-fired power plants in developing countries, and thereby credit to their own climate balance the CO_2 emissions they help to avoid.

The IPCC, the United Nations, the International Energy Agency, and coal-rich countries continue to place their hopes in CCS. They do this even though this technology is expensive and still in the early stages of development, and there are many uncertainties about the final storage of CO_2. Their attitude indicates a lack of confidence in other sources of energy. Global warming makes it an imperative to shift to a carbon-neutral supply of energy as soon as possible. Powerful interest groups, failing to apply innovative approaches and creative plans to work toward this end, continue to see coal, a fossil fuel, as a viable option—in the vague hope of making it environmentally compatible one day through new and expensive technologies that have yet to be developed.

THE FUTURE IS ELECTRICAL

The question of energy resources has taken a new direction with the progressive warming of the earth. Today it's no longer just a question about how much longer coal, oil, and gas reserves will last. The energy debate is also a climate debate. The future energy system must reliably guarantee global supply and produce as few greenhouse gas emissions as possible.

How is that supposed to work? One answer is to see the future powered by electricity, the form of energy with the highest quality. Electricity can be transported, easily controlled, and turned into other forms of energy—by engines turning electricity into motion, heat pumps turning it into heat, and lamps turning it into light. But the crucial factor in the climate debate is that electricity can be generated in a way that is climate-friendly—by hydroelectric power plants, wind parks, solar cells, solar thermal facilities, biogas, and wood-fired power plants, and indeed even nuclear power plants.

It is undisputed that the future belongs to clean, renewable energy sources. The numbers speak for themselves. The 2010 *Renewables Global Status Report* says that in Europe and the United States, year by year, more renewable energy capacity is being installed than fossil and nuclear energy. Wind power has boomed in recent years, with output increasing on average by nearly 30 percent annually since 2004. Chinese wind power capacity has caught up in giant steps and currently stands in second place, behind the United States and before Germany. Investments are also being made in wind power plants in North Africa, the Middle East, and South America. Photovoltaics are growing even faster, above all in Germany and Spain, with capacity increasing yearly by around 60 percent in recent years.

Biomass—wood, agricultural, and organic residential waste—is becoming more and more popular in Europe for generating power and heat. About a third of the electricity produced from biomass in industrial countries is in the United States, where even coal and gas power plants have been converted to biomass firing to some extent. Hydroelectric power will be attractive in the future, above all in emerging and developing countries. China has doubled its capacity in recent years, and Brazil, India, Russia, Turkey, and Vietnam are also counting on this clean energy. But the exploitation of renewable energies can also have social and ecological repercussions. Many hydroelectric power plant projects are highly controversial because they have involved the relocation of the population of entire cities, the best-known case being the Yangtze River project, which resettled 1.5 million people. The consequences of building dams are dire along rivers such as the Mekong, where farmers and fishermen in Cambodia and Vietnam depend on annual flooding.

But in spite of impressive growth figures, the share of renewable energy (wind, solar, biomass) in the global electricity supply today amounts to only 3 percent, and in total energy statistics it comes to only 0.7 percent. Even optimistic energy experts expect that renewable energy, not counting hydroelectric power, will not play a role in Europe's power mix until ten years from now, even though wind power is already making an essential contribution today in some countries like Germany and Denmark.

The International Energy Agency says that, apart from the transportation sector, the need for electricity in the next twenty-five years will rise more rapidly than that for oil, coal, and gas. Emerging and developing countries will experience the highest demand because their populations are growing quickly, and the desire for electronic luxury goods will increase as incomes rise. This demand will be augmented by the digital development of service operations, a high need for air conditioners, and growing power-dependent industrial production.

In contrast, the electricity market is already well established in industrialized countries, and their populations are growing slowly. Nevertheless, energy consumption will increase here as well because server systems are being expanded, the use of electric heat pumps for heating is booming, and the development of electric cars is expected to advance rapidly. The European Renewable Energy Council, in an optimistic scenario, calculates that electric vehicles could start becoming a frequent sight in traffic as early as 2020, although many engine manufacturers do not believe that electric cars will actually develop from a niche to a mass market product in the next ten years →p. 337.

Times are favorable for installing new and cleaner energy technologies. Industrialized countries expanded their energy infrastructures in the sixties and seventies, and many power plants in Europe and the United States are now reaching the end of their operating lives. Half of Europe's coal-fired power plants are more than thirty years old, and the situation is similar for gas-fired and nuclear power plants. Major investments in energy will be on the agenda in the next ten years. But the rules of the market economy will not easily allow new forms of energy generation to be installed because they are still not competitive when compared to amortized nuclear plants, which hardly incur any further costs. Apart from hydroelectric power, renewable energies need state support for now to survive against coal, oil, and gas as energy sources.

But renewable energies have gained ground in the last few years, thanks to technical progress, subsidies, and increased demand. Photovoltaic energy is a good example. Production costs for solar cells have dropped substantially in recent years, and engineers continue to look for new materials that will convert sunlight into electrical power more efficiently and that are cheaper to produce than current models.

Today, thin-film solar cells requiring much less silicon in manufacturing are already on the market. Perhaps the dye-sensitized cells developed by Swiss researcher Michael Grätzel, in which dye molecules capture rays of sunshine and convert them into electrical energy, will be the breakthrough. Visionaries are already imagining that window panes with Grätzel cells will produce electricity. Experts disagree as to whether photovoltaic power can be fed into the grid in Central Europe at market prices in ten years. The European Photovoltaic Industry Association (EPIA) estimates that photovoltaics will cover 12 percent of the demand for electricity in Europe by 2020.

Technically speaking, 100 percent of electricity in Europe could come from renewable energy sources by 2050, and indeed there is already a roadmap for this plan. The authors of this plan include consultants Price Waterhouse Coopers and the Institute for Climate Impact Research (PIK) in Potsdam. They foresee using a portfolio of energy sources: wind in the stormy North Sea; biogas, wood, and wind in the Baltic region and Eastern Europe; hydroelectric power in the mountains of Scandinavia and the Alps; and solar thermal power from the North African desert. Decentralized energy sources will have their place as well, whether they are photovoltaic modules on housetops and office buildings, city geothermal plants, or combined heating and power plants that supply electricity when needed

DEVELOPMENT OF TOTAL ENERGY CONSUMPTION WORLDWIDE BY ENERGY SOURCE

1990–2035, in billions of kilowatt-hours

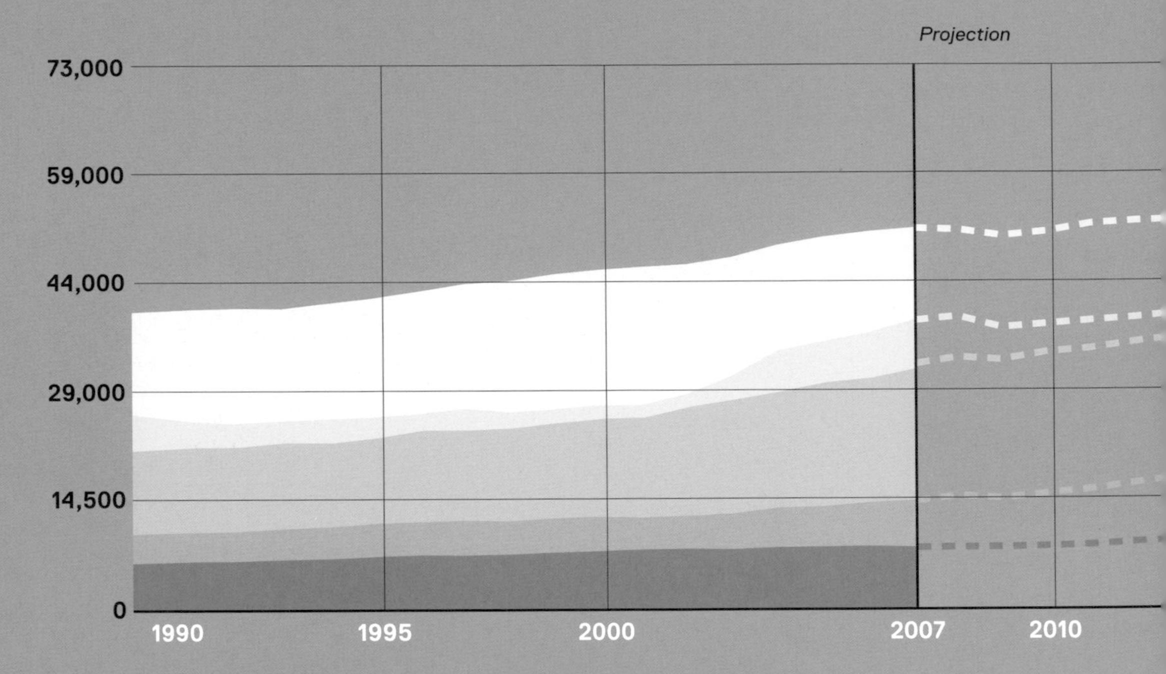

WORLDWIDE NET PRODUCTION OF ELECTRICITY BY ENERGY SOURCE

2007–2035, in trillions of kilowatt-hours

Net signifies that losses from power production, transmission, and consumption are deducted.

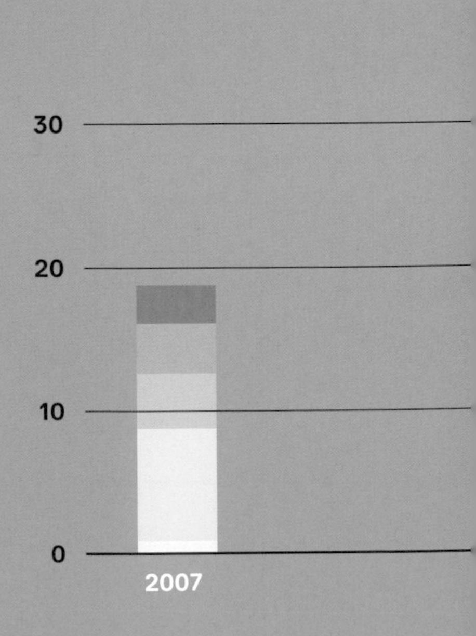

Source: World Energy Outlook, 2010

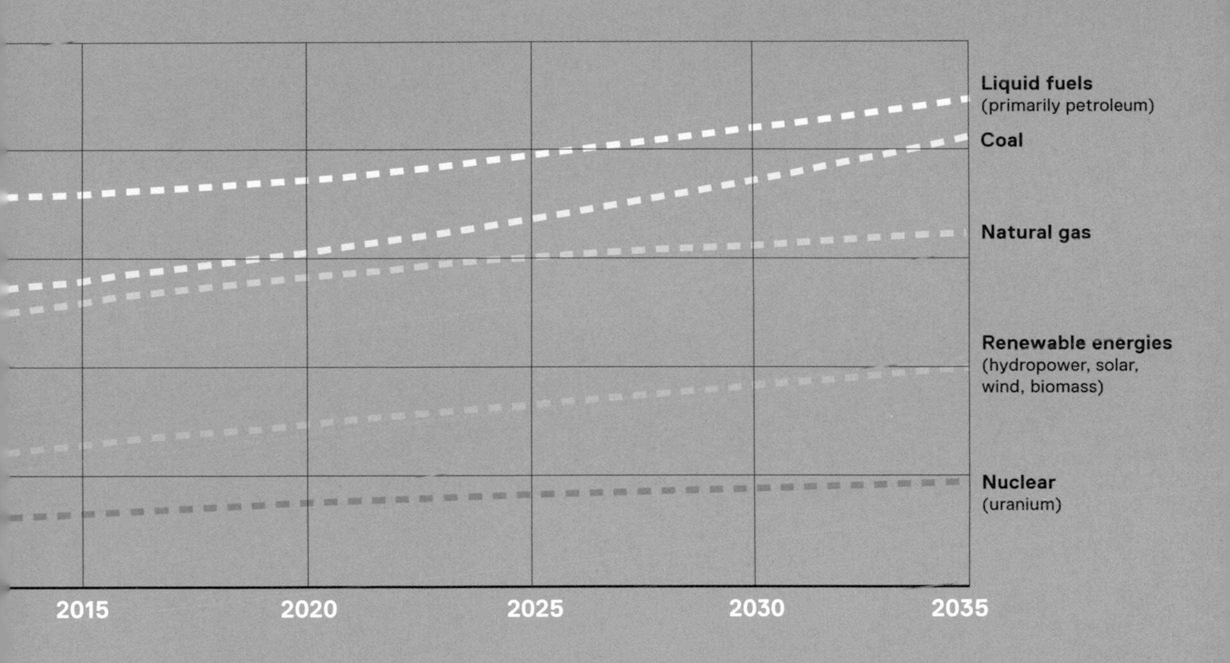
Liquid fuels
(primarily petroleum)
Coal
Natural gas
Renewable energies
(hydropower, solar,
wind, biomass)
Nuclear
(uranium)
2015
2020
2025
2030
2035

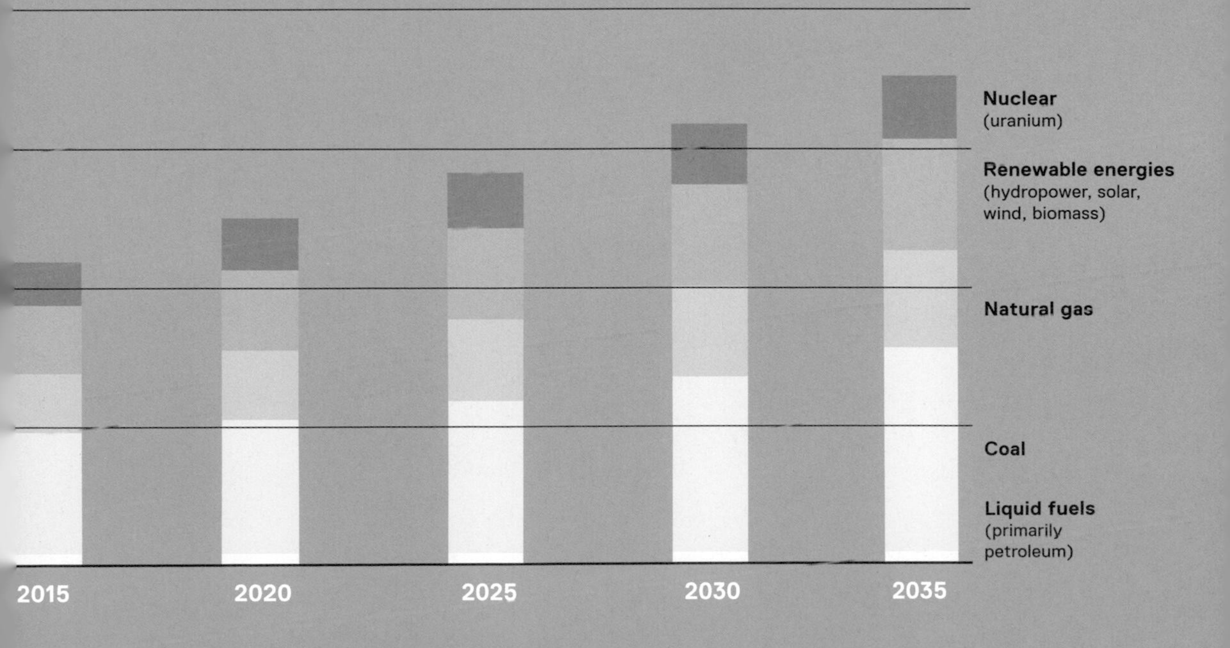
Nuclear
(uranium)
Renewable energies
(hydropower, solar,
wind, biomass)
Natural gas
Coal
Liquid fuels
(primarily
petroleum)
2015
2020
2025
2030
2035

and feed waste heat into a district heating network. The liberalization of electricity markets in Europe is forcing supply structures to open up that until now were in the hands of a few major utilities. Cities and towns are more and more interested in having their own, independent power supply.

The backbone of the future power system is supposed to be a direct current high-voltage network that transports electricity for thousands of kilometers without suffering much loss →p. 330. Existing conventional alternating current lines then distribute power over shorter distances into cities and towns. However, it will be a technical challenge to connect the high-voltage direct current (HVDC) transmission system to the conventional grid.

An additional sticking point will be to balance fluctuating power supply, from solar and wind sources in particular, in the grid itself. Power supply functions only when supply and demand are precisely adjusted to each other. Surplus power needs to be stored until needed. Today, surplus wind and nuclear power are already being used to pump water up into highly elevated reservoirs in pumped-storage hydroelectric power plants. Then, at peak demand times, water flowing back down through turbines generates electricity. There is a loss of some energy because the pumping process consumes more electricity than the turbines later generate. But there is an overall gain of energy that would have been lost had the excess energy from a renewable source not been stored.

But the future energy system will require even more storage. Engineers can imagine using plug-in hybrid cars, driven either with electricity or conventional fuel, or electric cars whose batteries function in an interim storage capacity →p. 352. This complex way of organizing electricity requires a creative system for distributing power, and indeed, scientists call this idea the "smart grid." This grid is supposed to extend into households, where intelligent meters, which have already been installed in many buildings, measure power consumption, and communicate directly with household appliances and energy utilities. For example, a refrigerator can be briefly switched off during peak power demand without the butter melting. Power suppliers and consumers enter into a completely new relationship. Utilities supply electricity, but can also withdraw the electricity stored in a car battery if it is needed. The power customer also becomes a power supplier.

This idea could become a viable option in all densely populated regions of the world. Savings in fossil fuels and the reduction of greenhouse gas emissions would be enormous in the long term. The cost of setting up this kind of infrastructure can't be reliably forecasted at this time. Ultimately it will be a question of political will whether renewable energies are subsidized in the long term, enabling production costs and risks for investors to go down.

In Europe, Germany and Spain in particular pay guaranteed prices, called feed-in tariffs, to make wind and solar power marketable. It is lesser known that fossil energies are also highly subsidized. The Global Subsidies Initiative project estimates that oil producers around the world receive support amounting to $100 billion of tax money each year. Admittedly, more and more public money is going into research, development, and demonstration plants promoting renewable energy, but half of the funding available for this kind of work still goes into fossil and nuclear projects. A relatively small share is used to work on improving energy efficiency. Compared to the years during the oil crises, today's share of public funding for energy research in the total volume available for research is quite modest, having decreased from 12 to 4 percent.

But the scenario for a sustainable energy system could be quite different. The lobby for nuclear energy is still strong, and the question of whether to construct new nuclear power plants has been reopened in recent years. The fact that nuclear energy produces very little CO_2 emissions is a major argument in the climate debate. Nuclear power plant operators see nuclear energy as a key solution for mitigating climate change. Although there is no final storage site anywhere for the disposal of radioactive waste, operators consider this problem to be technically solved. The International Energy Agency (IEA) also believes nuclear energy could be an option, next to renewable forms of energy, for partial replacement of fossil fuel power plants. Officially, more than fifty nuclear reactors are presently being built around the world, nearly twenty of them in China alone. Several are under construction in Russia, one is being built in Finland, and one in France. But critical energy experts doubt that the share of nuclear energy in power production around the world will go up in the future because numerous old reactors will have to be taken off the grid. The future of nuclear energy in Europe is uncertain. Experience with construction of the EPR reactor in Olkiluoto, Finland, is not very gratifying. The project has been delayed by three years and costs are already at least 55 percent over budget. In view of the issues of final storage, limited uranium deposits, and the danger that building more nuclear power plants also increases the threat that fissionable material could be abused for nuclear weapons (proliferation), it is doubtful whether the promotion of nuclear energy is politically desirable. Furthermore, skepticism regarding the safety of nuclear power has significantly increased again, at least in the Western world, after reactor core meltdowns at the Fukushima plant in Japan severely contaminated the surrounding area and seawater. A tsunami in March 2011, triggered by an earthquake with a magnitude of 9.0 on the Richter scale, caused the nuclear power plant's cooling system to fail entirely.

Experts in the energy sector believe it is unrealistic to completely write off fossil energy in coming decades. They assume that Asia's largest countries, China and India, will fuel most of their new power plants with coal. The energy sector believes that the only way to reduce CO_2 output from the combustion of coal is to invest in CCS (carbon capture and storage) technology, which separates out CO_2 and stores it underground →p.303. This technology is not new and has been used for a long time in the petroleum and gas industries, but it is far from being viable for use in coal-fired power plants. Like the problem with storing nuclear waste, reliable final storage sites have yet to be found underground that can hold billions of tons of carbon dioxide. Judging from 2009 green stimulus programs in the European Union, Australia, and the United States, a substantial share of research money is going into developing this technology.

In the short term, the greatest hope for mitigating climate change still lies in improving energy efficiency. From insulating buildings and using energy-saving lightbulbs, to operating more efficient electrical appliances and electric motors, even including innovative car engines, energy efficiency saves costs in manufacturing and for consumers. A United Nations report says that the global output of greenhouse gases could be reduced by 12 to 17 percent through efficiency measures alone. This would call for international agreement on the technical standards needed for key technologies to enable knowledge and experience to spread more quickly.

Vested interests in the energy sector are so different that it is difficult to see where developments will go. But one thing is certain: the global energy system will have to reorganize. Major energy companies prefer central structures that they can oversee and control. Nuclear power plant operators see the key to successful climate protection in nuclear energy, and coal-dependent countries are counting on their reserves and want to store CO_2 underground. The solar sector is hoping for long-term subsidies to remain competitive on the energy market. Patents play an important role in the expansion of renewable energy technology, such as photovoltaics, in developing countries. The United Nations Environment Programme (UNEP) says that most patents are held by industrialized countries, and a simple process for awarding licenses in poor countries has been a point of contention in international climate negotiations for years.

If we stick to the rules of the market economy alone, then we will probably not meet the long-term objectives of building a sustainable energy system based on renewable sources, and limiting the rise in average global temperature to 2 degrees Celsius (3.6°F). Political guidelines will be needed as orientation for the energy industry, as well as relevant economic stimuli.

The higher the speed, the more electricity is used. Chinese bullet train en route from Shanghai to Hangzhou.
Wang Dingchang/Corbis

Small is beautiful. G-Wiz electric car in the Chelsea district of London, United Kingdom. *Al Satterwhite/Corbis*

Pope Benedict XVI relies on an electrically powered vehicle—at least to move around within the small Vatican City State. *Global Electric Motorcars/Keystone*

The power grid is extensive, but not always reliable in Asian cities—overhead cables in Vietnam. *Moodboard/Corbis*

Access to electricity is a prerequisite to combating poverty in megacities. Overhead power cables in a favela in Sao Paulo, Brazil. *Noah Addis/Corbis*

The network of life in its modern version—power lines along the Savannah River in Georgia, USA. *Paul Souders/Corbis*

Growing demand for electricity in coming decades will call for enormous grid expansion; high-voltage power lines in Cape Cod, Massachusetts, USA. *Frank Siteman/Corbis*

Nuclear paradise lost — the nuclear power plant Isar II in Lower Bavaria, Germany, is scheduled to be shut down in 2022 as one of the last nuclear plants in Germany. *Paul Langrock/Keystone*

Visible signs of change—waters off the stormy eastern coast of Britain are a perfect location for large wind farms. *Rex Features/Dukas*

More and more people are taking electricity production into their own hands. The Japanese village of Ota is investing in a decentralized supply of energy. *Kyodo/Dukas*

Solar thermal power production needs a lot of space—the PS10 and PS20 tower plants in Sanlucar la Mayor near Seville in Andalusia, Spain. *Georg Gerster/Keystone*

Hydropower has it drawbacks, too. In China, 1.5 million people were relocated to make room for the Three Gorges Dam, the world's largest hydropower project, and the 660-kilometer lake behind it. *Wen Zhenxiao/Corbis*

DESERT POWER FOR EUROPE

Solar radiation in the desert is ideal for generating solar power. Solar energy from the Sahara is envisaged for making Europe's power supply significantly more climate-friendly.

Klaus Lanz

In 2005, experts at the German Aerospace Center (DLR) launched a bold idea to build huge solar power plants in North Africa's deserts, with a total capacity of up to 100,000 megawatts, able to supply enough electricity to meet regional needs by 2050, and even supply up to 15 percent of Europe's demand for electricity. The DLR argued that the technologies needed for generating electricity and transmitting it long distances to Europe's urban centers were reliable and fully tested. The only thing needed to realize the project, called Desertec, was political support and determined investors.

A MATTER OF TECHNOLOGY

The problem of transmitting large amounts of electricity to faraway consumers, one of the project's basic requirements, indeed no longer poses a technical problem. The breakthrough was achieved through high-voltage direct current (HVDC) transmission, a technology that loses only about 3 percent of generated energy per 1,000 kilometers. This allows ABB, a Swiss-Swedish technology group, to ensure that China's hydropower can be delivered to consumers throughout the country, for instance. Likewise, Siemens, a provider of energy systems, made electricity from the Australian mainland available to consumers on the island of Tasmania.

The DLR pointed out that solar thermal power plants, also called concentrating solar power plants (CSP), have been in use for a long time and can rely on mature technology. The first such plant went into operation in California's Mojave Desert back in 1981. It was equipped with parabolic trough collectors, as are most solar thermal power plants built to date. CSP plants work a lot like other thermal power stations, whether fueled by nuclear energy, coal, or gas, since they run turbines to produce electricity. Parabolic mirrors focus solar radiation onto receiver tubes. Concentrated sunlight heats thermal oil flowing through the tubes to 400 degrees Celsius (752°F). Heat from the oil is transferred to water through a heat exchanger, generating the steam needed to run turbines that produce electricity. CSP power plants are able to supply electricity twenty-four hours a day as they can store collected heat in molten salt reservoirs, for instance, and use the heat at night to continue producing steam.

Solar thermal power is completely different from photovoltaic systems that convert sunlight directly into electricity. CSP power plants are more like conventional power plants in that they use steam turbines and are not suitable for decentralized use. They require a great deal of space—depending on the intensity of the sun, solar collectors covering an area of 25,000 to 40,000 square meters are needed per megawatt (MW) of installed capacity. In southern Spain, a 100 MW power plant with enough capacity to supply electricity to 400,000 people covers an area measuring roughly two kilometers in both length and width.

These plants need such extraordinarily large tracts of land because their electricity yield is low. Only a small portion of the solar radiation is actually converted into electricity. Over 70 percent is lost in various stages of conversion: solar radiation into oil heat, oil heat into steam, steam into electricity. Moreover, conventional CSP power plants require large amounts of cooling water, which means they must be sited in arid regions close to the sea. Only innovative, water-saving cooling technologies, such as the Heller system, for example, make it possible to generate solar thermal power in completely arid regions, but these cooling technologies are much more complex and very expensive.

Solar power plants operating on the basis of a Stirling engine do not need water either. In this system, solar radiation is directed to the heat engine where the solar heat is first converted into mechanical energy and then into electricity. These systems generate considerably more electricity per square meter of reflecting surface than solar thermal power plants that rely on steam turbines. More recent studies have also brought photovoltaic technology for desert power plants into play. When sunlight is highly concentrated through lenses and focused on high-performance solar cells, electricity yield can be nearly doubled and the cost of photovoltaic electricity substantially lowered. There are many solar technologies being developed and tested throughout the world today. It remains to be seen which one will turn out to be best suited to desert conditions.

The choice of technology will also influence future energy management structures. While photovoltaics and Stirling systems are ideal for decentralized applications, solar thermal power generation requires large central plants. The first can be deployed by small investors, whether they are homeowners, businesses, or municipalities, but the latter depends on the involvement of multinational energy groups such as RWE, E.ON or EdF. Energy experts who are critical of Desertec maintain that the main reason the project is focusing exclusively on solar thermal power is to ensure that the electricity suppliers involved keep their hold on the market. They also point out that it would be far cheaper in North Africa to generate electricity from wind power than from solar energy—yet wind power does not play a role in the Desertec project.

INVESTORS WANTED

Uncertainty over which technology is best suited for the project is not Desertec's biggest obstacle however. A megaproject of this kind needs substantial investments of up to 400 billion euros by 2050, including 350 billion euros for power plants alone. For this reason, the Desertec Industrial Initiative (DII), a consortium of mostly German companies, was founded in 2009 to leverage the vision of green energy from the desert at the political, technological, and financial levels. The consortium's partners include equipment manufacturers Siemens and ABB, electricity suppliers RWE and E.ON, solar enterprises Solar Millennium, SolarWorld, and Schott Solar, as well as the financial service companies Munich Re and Deutsche Bank.

But for the time being, these groups are still holding off putting their own money into solar power plants. Indeed, it is only when profits are certain that the development of solar power plants progresses at great speed. A case in point is the Andasol 3 solar thermal power plant that RWE is currently building in Spain at a cost of 400 million euros. Although electricity generated at this plant will cost RWE around 25 cents per kilowatt-hour, the Spanish government is subsidizing solar thermal electricity at 27 cents per kilowatt-hour, and RWE will earn another 5 cents on each kilowatt-hour it sells.

Businesses involved in the Desertec project wonder why this does not work in Germany, where the Renewable Energy Sources Act has earmarked billions for solar energy subsidies. To make the project profitable for investors, they are calling for electricity produced in the desert and exported to Germany to be subsidized as it is in Spain. Currently this does not seem likely to happen. Mention of the Desertec project in the German government's 2010 energy plan was made only after the consortium did substantial lobbying.

Working toward a climate-friendly future—the Andasol 2 solar thermal parabolic trough power plant in Spain has been an important step; today it has a capacity of 50 megawatts. *Paul Langrock/Keystone*

Energy from the desert can count on more political support in France. The French government launched the Transgreen initiative in July of 2010, a consortium put together to build a grid of high-capacity transmission lines between North Africa and Europe. Transgreen, now called Medgrid, expects to install several submarine cables by 2020, enabling the transmission of 5 gigawatts of North African solar power to Europe, a volume equivalent to the capacity of three nuclear power plants.

Medgrid does point out that it is focusing on creating the necessary intercontinental transmission capacity and is in no way competing with Desertec. At the same time, the declared aim of the consortium is to develop foreign markets for European energy technologies. Medgrid's aims are clearly consistent in this respect with French president Nicolas Sarkozy's energy policies. After decades of giving nuclear power top priority, France now plans on breaking into and developing its own solar energy industry, otherwise belittled and neglected for so long, and the French government is making use of its close ties to Francophone North African states. Desertec, a consortium dominated by German companies, lacks such connections. Although the Medgrid consortium consists predominantly of French,

Pioneer work — the world's first solar thermal power plant was built in the Mojave Desert in California and produced electricity until 1999. *Georg Gerster / Keystone*

Spanish, and Italian electricity and technology enterprises, the Moroccan Agency for Solar Energy and Syrian and Egyptian businesses are also involved.

THE PERSPECTIVE FROM HOST COUNTRIES

States in North Africa and the Middle East quite naturally consider the idea of generating electricity in their deserts from another perspective. In principle they are not opposed to exporting solar power generated in their desert regions to wealthy European countries at some point in the future in exchange for good money. But their priorities lie with providing their own populations and industries with enough electricity to enable modernization.

Morocco, for instance, a constitutional monarchy in the northwestern part of Africa, lags far behind in its economic development and depends almost completely on imported oil for its power supply. To escape this dependency, the monarchy plans to use wind and solar power to generate roughly 42 percent of the country's electricity by 2020—representing a threefold increase in its power plant capacity compared to 2008. Half of this electricity is to be produced by solar thermal power plants at five designated sites; the first 500-megawatt power station in Ouarzazate is expected to be completed by 2015. Morocco plans on investing a total of 6.6 billion euros of its own money by 2020.

Morocco understandably wants to keep open its option to export solar power to Europe. It is dependent on technical and financial support from European countries to bridge the distance across the Mediterranean. Morocco is waiting to see whether its partners will be German or French or perhaps both; for now the monarchy is involved in Desertec with ONA, a state-owned industrial group, and in Medgrid with ONE, the Moroccan national electricity agency. Moroccans are also working with other partners. Early in 2010, an agreement was signed with the Japanese government to build Africa's largest photovoltaic plant in Assa-Zag. This 1-megawatt solar power plant may be just a small start, but with enormous amounts of sun and wind at its disposal, Morocco could soon become a key player in international energy politics.

Morocco is already at the focus of European plans for accessing solar power. The industry's interest is huge and the Moroccan government received fifty international bids to build the first CSP plant in Ouarzazate. Technology businesses in the Desertec consortium see Ouarzazate as a milestone in the realization of their ideas and perceive it as a model for future power plants.

NEW CONFIGURATIONS

Recent political upheavals in North African countries will also have implications for Europe's desert power aspirations, but it is impossible to know to what extent. Egypt, Tunisia, and Libya are in a precarious period of transition and no one knows whether the political situations in Algeria or Morocco will remain stable over time. Not one of the countries in which solar power plants are to be built has a political system that corresponds to the European idea of democracy—and there is no promise of long-term investment safety. Whether electricity generated in the desert will one day flow to Europe depends on the willingness of big investors to take risks.

Europe itself is also far from satisfying the conditions for shifting to sustainable energy systems. There is no agreement on introducing EU-wide feed-in tariffs, and international trade with green energy is nonexistent. Both are important prerequisites for ensuring the success of the Desertec project. Many countries still pursue isolationist energy policies, preventing the rapid breakthrough of solar and wind energy. And while there is a lot of talk about constructing an intercontinental supergrid between Africa and Europe, not one transmission line crosses the Pyrenees. There is a general shortage of transmission capacity between countries throughout Europe, and indeed there are bottlenecks at all borders. According to former UNEP executive director Klaus Toepfer, "Cross-border transmission capacities between national energy grids in Europe were not built to allow the expansion of an efficient European energy market."

Desertec's main goal, to improve the CO_2 balance of Europe's electricity supply by importing electricity generated in the Sahara, does seem to be technically feasible today. But for now, Arab revolutions have postponed the ambitious plan until some time in the distant future. The Desertec consortium will have to come to terms with new partners and changed systems, whether they are democratically elected governments or other power structures.

BILLIONS OF CARS

Mobility can be a very emotional issue if talk is about saving energy and mitigating climate change. Our affluent society is used to driving or flying from A to B in the shortest possible time. Globalization would not be possible without a reliable global traffic system.

No less than a quarter of global carbon dioxide emissions comes from traffic and transportation (including air travel); this share is approximately one-third in Europe. Most of the fuel used for transportation is produced from fossil resources, namely oil, with road traffic taking up a good 40 percent of global oil consumption. This percentage will continue to increase because in the future electricity will be generated less from oil and more from energy sources such as wind, natural gas, and coal.

Therefore it seems obvious to start effectively reducing greenhouse gases in the transportation sector, although the sheer growth in the number of cars around the world is an obstacle to progress. Deborah Gordon and Daniel Sperling, two reputable American experts in mobility, have said that there seem to be more parking lots than streets in growing megacities like Bangkok and Moscow. There will soon be more than one billion vehicles on the planet, and experts believe this figure will double to two billion in twenty years. Although growth in the number of cars has been only around 1 to 2 percent in the United States, Western Europe, and Japan for the past ten years, the growth rate in emerging countries is rising from year to year. Growing income creates the desire for a privately owned car, which is often seen as a sign of economic success.

The dimensions here are unimaginable. Some 2.4 billion people live in China and India. Approximately 35 of every 1,000 people in China today own a motorized vehicle (including trucks and buses); in the United States, more than 800 of every 1,000 Americans own a vehicle. If we assume that the need for individual mobility increases to the same degree in developing and emerging countries as in wealthy states, then there will be about four billion vehicles on the planet someday. This calculation is based on population figures of around seven billion people today and an average vehicle density of 600 per 1,000 people in industrialized countries. This is a game of numbers, but it does show that making engines efficient and clean will not be enough to mitigate climate change. Controlling growth in the number of cars on the road and finding ideas for making mobility without car ownership attractive will have a decisive impact on future demand for energy.

The automobile industry offers a whole range of solutions for the clean car of the future, but it is difficult to foresee what technology will finally prevail in coming decades. Electric cars would doubtless be the most elegant solution for reducing greenhouse gas emissions and air pollution if we could assume that the electricity came exclusively from renewable energy sources like the wind and the sun. An electric drive converts 75 percent of the electrical energy from a power outlet into mechanical energy, in contrast to the internal combustion engine, which has an efficiency of only about 20 to 25 percent. This figure goes down to roughly 18 percent when the entire energy chain is taken into consideration, from the extraction of oil through transportation and distribution up to combustion in the engine.

Plans for electrically driven vehicles are ambitious. Ten years from now, every tenth car in Denmark is supposed to be powered by wind energy. Volkswagen, the German auto group, has announced an ambitious strategy for manufacturing electric cars and hybrid vehicles. Three percent of the entire range of VW cars for sale in 2018 is supposed to be powered by electricity. Some very promising electric cars from Volvo, Renault, Nissan, Volkswagen, Toyota, Peugeot, Citroën, and Ford will come on the market in the next few years. A thousand Smart Electric Drive cars are currently being test driven in Paris, Rome, Milan, and Manhattan.

But the electric car revolution could take place in Asia. China plans to produce one million electric cars by 2020. A test program is already running in which electric vehicles are being promoted in twenty-five major cities like Shanghai and Shenzhen—with big financial support from the government. At its center is the young enterprise Build Your Dream (BYD), which was founded only as recently as 1995. The professional world took note of the first Chinese plug-in hybrid, which has a small internal combustion engine, a complete electric drive system, and a storage battery. The battery is charged internally either by a generator driven by the internal combustion engine or by so-called recovery energy generated during braking of the vehicle. The hybrid can also be attached to a power outlet. Fully

WORLDWIDE DISTRIBUTION OF CARS BY COUNTRYWIDE INCOME – in percent

Income is defined by the World Bank as the per capita gross domestic product.

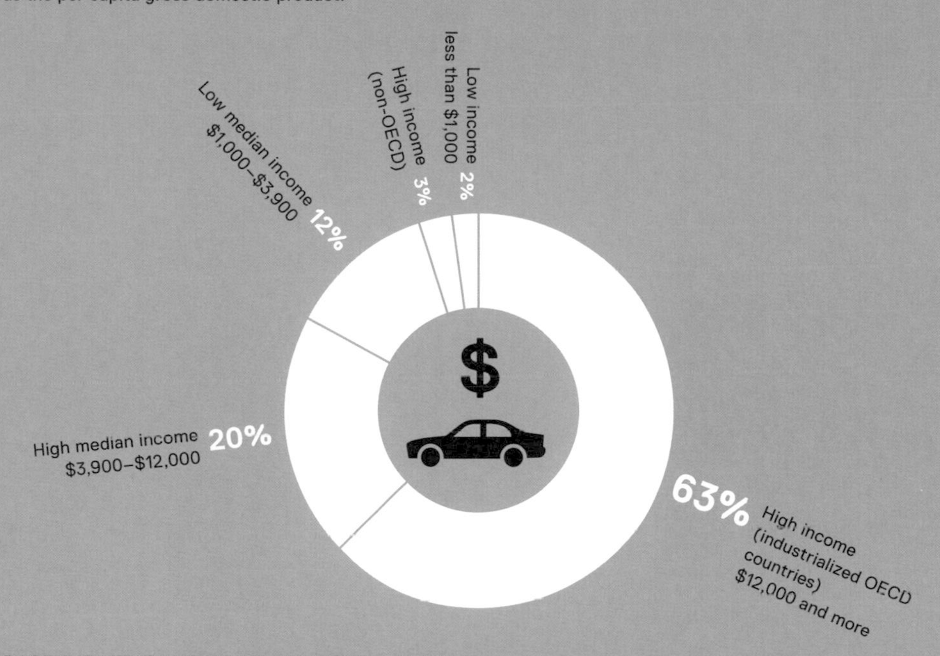

DISTRIBUTION OF VEHICLES BY REGION

in percent (including buses, minibuses, and vans)

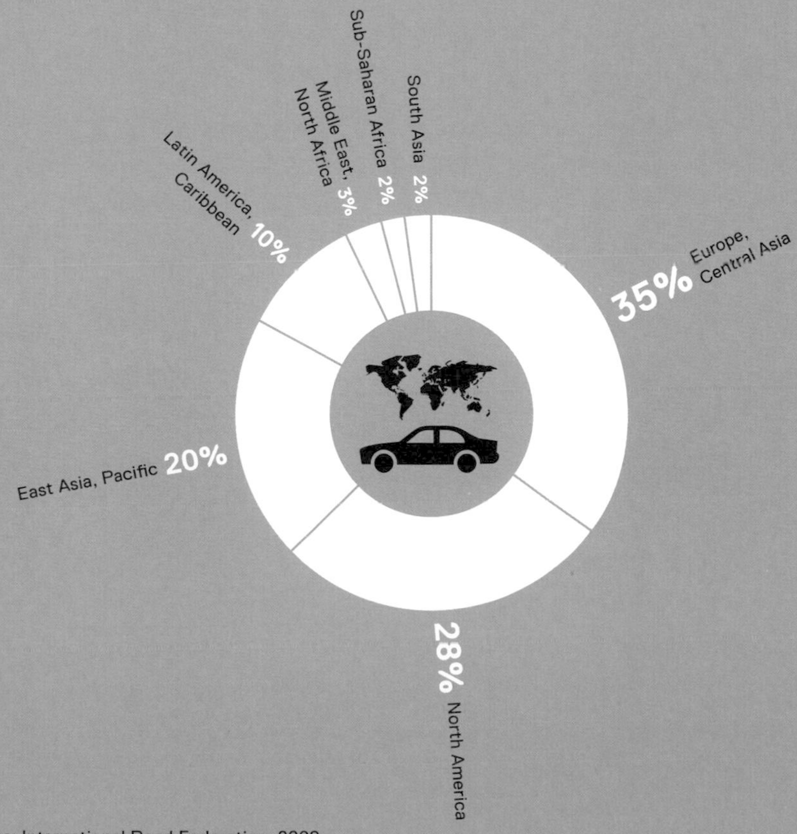

Source: International Road Federation, 2009

There are more than a billion vehicles on Earth. Experts predict that in twenty years there will be two billion.

charged, the car has an operating distance of 110 kilometers when it is running on electricity alone. BYD recently announced plans to manufacture a purely electric car that would travel 290 kilometers. This Chinese enterprise has the advantage that it did not start out in the automobile industry, but as a developer of batteries.

The battery, at the heart of electric cars, also poses a challenge for engineers. Its energy density is around 3 to 4 percent of that for a tank of gasoline. As a rule, 100 kilograms of battery are needed for a vehicle weighing one ton to travel 100 kilometers. This not only makes a car heavy, it also makes manufacturing much more expensive. Furthermore, charging the car at a power outlet takes several hours. Charging up at a rest stop on the autobahn is unthinkable unless the idea of exchanging batteries at a power station takes hold.

But there are clear signs of progress, starting with the development of the lithium-ion battery. In any case, new ideas will be needed to lower costs, optimize the charging and discharging processes, and reduce weight. Although Toyota uses nickel-metal hydride (NiMH) batteries in its Prius hybrid models, it seems that lithium will play an important role in the future. Swiss researchers at the Swiss Federal Laboratories for Materials Science and Technology (Empa) can attest that lithium-ion batteries have good environmental performance. Manufacturing, maintenance, and waste disposal make up about 15 percent of the environmental impact of an electric car. Charging the battery is of more consequence. A medium-sized car powered by electricity from the European power mix (30 percent from coal, 20 percent from gas-fired power plants) corresponds approximately to a car with a gasoline engine that consumes between three to four liters for 100 kilometers.

Even if there are significant advances in battery technology, it will be decades before electric cars become common in the global fleet of vehicles. Gasoline and diesel will probably remain the most important fuels in coming decades. Internal combustion engines are cleaner now, and there is still room for improvement in weight, aerodynamics, and engine performance. Engineers also believe that small cars consuming less than 2.5 liters of gasoline per 100 kilometers can become a reality.

The United States has adopted a program requiring car manufacturers to lower fuel consumption and greenhouse gas emissions in the next generation of cars by around 5 percent every year by 2016. The

European automobile industry will also have to take steps soon because its fleet of new cars is supposed to emit less than 130 grams of CO_2 per kilometer starting in 2015. Average emissions from new cars in Europe today are approximately 155 grams of CO_2 per kilometer, and in the United States they are about 215 grams.

Another alternative for significantly lowering traffic emissions is the natural gas car. Studies show that natural gas and hybrid vehicles emit up to 25 percent less CO_2 compared to gasoline-engine cars. Hybrids emit less CO_2 in city traffic; natural gas cars less on the autobahn. Natural gas engines can also be fueled with biogas from organic wastes, liquid manure, or waste wood, making them CO_2-neutral.

The highly touted fuel cell, on the other hand, has not made a breakthrough so far. Its technology is based, in simplified terms, on the reaction between hydrogen and oxygen, and water is its waste product. The chemical process generates electrical current. Although there are several prototypes of fuel cells, their development seems to be faltering. The fuel cell seems to promise efficiency and CO_2-free technology, but it has not been tested long enough for mass production. Manufacturing costs are high and there is a problem with infrastructure. A great deal of energy is needed to store hydrogen fuel in dense gaseous form or at minus 250 degrees Celsius (-418°F) in liquid form. A hydrogen tank must be many times larger than a gasoline tank in a vehicle for it to obtain the same operating distance. Critics of hydrogen technologies point out the unnecessarily high energy loss involved in first generating hydrogen through electrolysis with power from renewable sources, and then later burning the hydrogen in a fuel cell to produce electricity again. This cumbersome and inefficient chain of conversions does not apply to electric cars.

A further option for reducing CO_2 emissions is to use agrofuel →p.392. This is actually not a new idea. In the 1920s and 1930s, the pioneering era of automobile manufacturing, engineers tried adding fermented alcohol to gasoline. Today, half the fuel for vehicles in Brazil is bioethanol made generally from sugar beets, and practically every new car in Brazil is capable of running on bioethanol alone, on gasoline, or on a combination of the two. The biggest producer of bioethanol is the United States, where this fuel is produced almost exclusively from corn starch. The share of biodiesel is significantly smaller in the global supply of agrofuels. Here the EU holds the lion's share of nearly 50 percent, although it is also manufactured in the U.S., India, Argentina, Colombia, and Indonesia. Palm oil, soybeans, rapeseed, and sunflower seed oil serve as raw materials. Poorer countries like Kenya, where there are large areas of dryland, are interested in the oils of the jatropha plant, which is resistant and can be cultivated in less fertile fields.

But agrofuels are controversial. The loss of cropland for food security is one of the main arguments against first generation agrofuels. The focus is increasingly on a second generation that does not compete with food production. Lignocellulose, basically the structural material of all terrestrial plants, is used as raw material for fuel production; it includes grasses, straw, mash, and wood. Algae could also play a role as a raw material in the future. In spite of the uncertain development of agrofuels, many states are counting on this option for making car, train, and air traffic less harmful for the climate, among them Brazil and the U.S. The EU also intends to promote agrofuels and has already introduced strict ecological standards.

In spite of innovations that hold promise, we cannot avoid taking a closer look at our need for mobility, especially because looking at the energy consumption and emissions involved in driving a car does not reveal the full picture. The manufacturing of vehicles is also an energy-intensive process, and annual global production can generate as much CO_2 as air travel does in a year. This is a considerable amount if we note that a car in Germany, for instance, is parked for an average of twenty-three hours a day.

There are various options for changing our ways of being mobile. If public transportation and car sharing in cities are attractive, then people can more easily do without owning a car. London, Stockholm, and Singapore have been able to relieve traffic in their city centers with road pricing. Vancouver massively lowered the share of people traveling into the city by car by expanding various pedestrian zones and painting designated bicycle lanes on ten main streets. Paris offers bicycles for travel within the city at a low price.

Urban and traffic planners are called upon and have the power to develop new plans for mobility and substantially decrease energy consumption and CO_2 emissions.

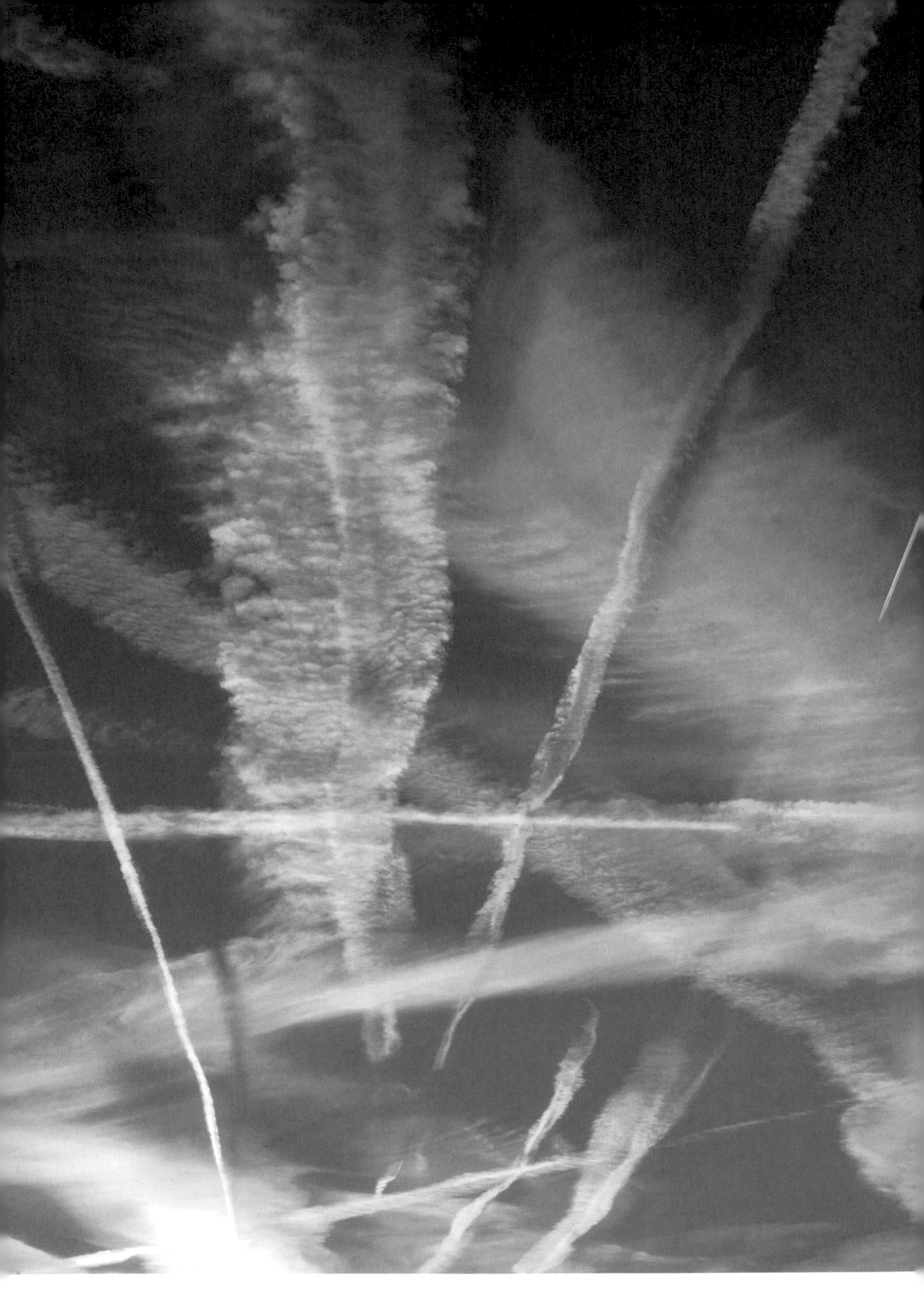

 Busy skies — tour operators and airlines talk of reducing greenhouse gas emissions in the future. *Lothar Stengel*

Mitigating climate change calls for new perspectives in mobility—evening traffic in Manhattan, New York. *Danny Lehman/Corbis*

Urban planning is a major challenge for many rapidly growing cities in developing countries—congestion in Lagos, Nigeria's largest city. *George Esiri/Reuters*

Are rickshaws a nuisance to traffic or a contribution to sustainable mobility? A rickshaw traffic jam in the center of Dhaka, the capital of Bangladesh. *Frédéric Soltan/Corbis*

More and more vehicles on the road in Asia—like in Nepal's capitol Kathmandu, urban infrastructure in many cities is unable to cope with the increase in traffic. *Narendra Shrestha/Keystone*

No room for individual cars in Karachi, Pakistan, where buses inch their way through Saddar, the city's central business district. *Akhtar Soomro/Keystone*

The easier it is for pedestrians to make their way around a city, the less cars are used. A business district in Hong Kong, one of the world's leading financial centers. *Xinhua/Landov/Keystone*

The future is already here — monorail service in downtown Miami, USA. *Modrow/laif/Keystone*

No matter the weather, bicycles can be an important means of individual transportation in cities. Copenhagen, Denmark. *David McLain/Aurora/laif/Keystone*

A well-designed public transportation system reduces traffic on the streets—the Metro in Saint Petersburg, Russia. *Sylvain Grandadam/Keystone*

Moving forward together — combining different types of transportation improves urban traffic flow. Florence, Italy. *Michael S. Yamashita/Corbis*

TAXI

VEHICLE TO GRID

Tomorrow's power sources, wind and sun, generate electricity that is intermittent. To base a reliable and continuous energy supply on these sources, the electricity they produce needs to be effectively stored during peak times. Batteries in electric cars could play an important role.

Klaus Lanz

In the near future, solar and wind farms will supply a significant share of our electricity. One drawback is that sunlight and wind are dependent on weather and seasonal conditions, making their production of electricity intermittent. Energy storage facilities could supply the power needed during periods of low yield. Excess electricity generated at peak production times could be stored and released whenever demand increased.

Electrical storage facilities of this kind already exist—indeed they belong to the business concept of Switzerland's energy sector. Whenever French nuclear power plants or coal-fired power plants in Germany produce more electricity at night than is needed, the Swiss buy the excess electricity and use it to pump water up into reservoirs in the mountains. The next day, when demand increases again, pumped storage hydropower plants release exactly the amount of energy needed—at much higher market rates.

The problem is that there is not enough storage capacity available yet to regulate fluctuations in consumption. And fluctuations in wind and solar energy production will further aggravate the shortage of storage capacity, and it will become increasingly difficult to keep supply voltage stable and ensure that electricity is available twenty-four hours a day. In the future, power grids therefore need to be managed in a much more flexible and intelligent way, in an operational mode referred to as a smart grid.

THE NEW ROLE OF THE ELECTRIC CAR

Electric car batteries could help stabilize energy supply by acting as a cache for surplus power. Parked cars would have to be connected to the grid so that their batteries could be charged and discharged (to a degree) whenever required. Cars are in parking areas or garages for an average of about twenty-three hours every day. In theory, if researchers found a way of integrating car batteries to the grid during this time, this would create an enormous cache for electricity.

Leading electricity groups, such as RWE in Germany, EdF in France, and Siemens (as a components manufacturer), are convinced that this idea is technically feasible and expect this market to generate billions in revenue. The idea foresees the installation of computer-monitored charging and discharging stations in public parking lots and private garages. Owners of electric cars would receive a credit for charging their batteries with cheap excess electricity at night and for supplying the stored energy to the grid during peak daytime usage hours when electricity is much more expensive.

This technology would help wind energy and solar power make a breakthrough, adding to the incentive to buy an electric car and participate in the vehicle-to-grid system. The energy industry argues that the finely tuned charging and discharging of car batteries would help make full use of electricity produced during peak periods of strong wind. Excess energy could be stored in this way and then released back into the grid during slack periods.

SWITCHING FROM OIL TO ELECTRICITY

The question is whether it really makes sense to switch to electrically powered cars just to create new power storage capacity. The cost and work involved in installing a completely new and extensive infrastructure of charging stations and consumption meters is immense. There are of course other ways of creating caches for surplus power; for instance, battery energy storage systems maintained by power companies. ABB is currently working on further development of this type of storage system, used for decades in West Berlin to stabilize the city's isolated energy grid.

Yet large energy suppliers such as RWE and EdF continue to put a great deal of effort into propagating the vehicle-to-grid system. Jürgen Grossmann, CEO of RWE, has declared, "We need as many e-cars as possible on the roads. This is our goal." Siemens CEO Peter Löscher shares this view: "E-mobility is a powerful solution," he has said, and expects the infrastructure business in particular to generate "a multibillion-euro market."

From the point of view of energy companies, their commitment to electric cars and the vehicle-to-grid system is self-evident, allowing them after all to lay claim to an entirely new business segment. If all private transportation did in fact make the switch to electrically powered vehicles, it would represent a tremendous restructuring of energy markets. Electricity suppliers would be able to significantly expand their market potential, while the oil industry would lose its most important economic mainstay.

At first sight, this might seem like a good thing. Electric cars are quiet and do not release any exhaust fumes or carbon dioxide. The drawback is that demand for electricity would rapidly increase because individual transportation is energy-intensive. In real numbers, if all cars in Germany were powered by electricity, for instance, RWE estimates that demand would increase by 88 billion kilowatt-hours per year. This figure is about 85 percent of the total power consumption of German households (heating excluded), and would nearly double the private use of electricity for lighting and powering—equivalent to an additional turnover for energy companies of 2.6 billion euros.

Where are we supposed to get the additional energy needed for electrically powered, private transportation? Part of it could come from optimizing the use of existing power plant capacity, especially by installing more efficient storage systems. In fact, energy suppliers are focusing all their energy on the vehicle-to-grid idea, but they are less vocal in communicating that this would hardly make available enough electricity to meet additional demand. In other words, new power plants would have to be built for electric cars while old power plants would continue to run for a long time to come. There is no way this can be thought to help mitigate climate change. If today's energy mix in Europe remains as it is, then for decades electric cars would be responsible for emitting just as much CO_2 as cars with high-performance internal combustion engines. The costs involved and the overall CO_2 balance of electricity therefore speak against making a complete switch to electrified private transportation in coming decades.

This is why independent experts warn against setting up a comprehensive network of charging and discharging stations and making an irreversible commitment to electrically powered individual transport. More importantly, it is doubtful whether in the long term—regardless of the type of energy used—we can sustain a system of transportation that is 80 percent dependent on cars. On a global scale, this would mean a billion new cars by 2050—more than double the number on Earth today. Traffic planners today

The introduction of a congestion charge in London has triggered a boom in electric and hybrid cars because they are exempt from paying the fee; owners can save up to 2,000 pounds per year. *Martin Goddard/Corbis*

call for sustainable mobility systems in which privately owned cars no longer play a dominant role, especially in urban areas.

By portraying a switch to electric cars as a clean and responsible idea for the future, energy suppliers are promoting the globalization of Western-style private transportation. Energy needs for transportation in this scenario would continue to grow massively. The assertion that the vehicle-to-grid system is an ideal environmental model is very questionable. In reality, benefits related to the climate would indeed be slight, and it is more the power industry that would stand to profit immensely from a restructuring of global energy markets in their favor.

SIMPLER SOLUTIONS

There is no doubt that electric cars will be an important component in the transportation mix of the future, but they are not the only solution. There are simpler and smarter alternatives to smart grid technology. Like night storage heaters and water boilers, car batteries can be charged with off-peak electricity, which is abundant and cheap. They can even be charged during the daytime without putting too great a strain on the grid. Swiss engineers have developed a compact measuring unit for electric cars that determines the frequency and voltage of electrical supply and interrupts the charging process whenever there is not enough electricity available. In a manageable way, this system could succeed in counterbalancing the fluctuations in supply that are so typical of wind and solar energy.

A business has every right to fight for its share of the market. But we should know that if RWE and EdF are granted the subsidies they are seeking for the multibillion development of a vehicle-to-grid infrastructure, they will be setting a course that would effectively prevent the development of other technologies and sustainable mobility concepts. By 2025, the not so far future, 60 percent of the world's population will be living in large urban centers. Motorized private transportation has no place in such megacities, not only because it consumes energy but also because of the amount of surface area it takes to build wide streets and parking facilities.

THE CHALLENGE OF MEGACITIES

Cities are the engines of the economy. Places with strong economic performance are where the most energy is consumed, and where the most greenhouse gases end up in the atmosphere. Seventy percent of global emissions come from urban areas. Half the world's population now lives in urban settlements. According to the UN Agency for Human Settlements (UN-HABITAT), in forty years it is likely to be six out of every ten people. The greatest growth is expected in developing and emerging countries. What the Western world already has behind it is about to happen there—an urbanization and industrialization revolution. Emissions are expected to rise significantly in the process, especially in China and India, while in the West the curve for emissions will more likely flatten out.

There are already more than 190 cities in Asia with more than one million inhabitants, and more than ten million people already live in each of eleven metropolitan complexes in Asia. The five largest megacities in China, each with more than ten million inhabitants, are responsible for 15 percent of the national economy. But demographers believe the biggest growth in the future will be in new medium-sized cities with 100,000 to 200,000 inhabitants.

City planning in the future will therefore have great impact on the global CO_2 balance. Many countries now recognize that there are real opportunities for designing cities that are environmentally friendly. China is investing a great deal of money in developing sustainable city plans. On the island of Chongming near Shanghai, the first "CO_2 neutral" city, Dongtan, is planned for a population of 10,000, with a tunnel connecting the island to the megacity. Supporters of this project are hoping to set a model for China's future urban planning. Half a dozen such "ecocities" are on the drawing boards of urban building authorities. The Chinese government has stated that this is China's last chance to turn urbanization in the right direction. Many cities are suffering from the chaotic building of the eighties that saw manufacturing zones sprawl endlessly without residential or shopping areas.

The Dongtan planners are working toward an urban framework that allows inhabitants to reach shopping centers and public transportation within a ten-minute walking distance. Only vehicles with no

CO_2 emissions such as fuel cell or electric cars will be permitted in the city. Buildings are insulated, and state-of-the-art lighting and household appliances will be installed, consuming low amounts of energy. Photovoltaics, wind, and biomass technology will generate electricity, and solar panels will be responsible for hot water. Recycled wastewater will flush the toilets and provide the water for vegetable and fruit gardens integrated into "high-rise greenhouses" between office buildings. Dongtan is supposed to be an open-air laboratory for China's future city planning. Planners aren't willing to talk about costs because they say it isn't yet possible to make comparisons with conventionally built cities.

Another experiment is underway in the United Arab Emirates. Masdar, an ecocity for 50,000 people, is supposed to come into being about seventeen kilometers northeast of Abu Dhabi. Inhabitants are expected to need only a fraction of the energy used in traditional cities. Planners are looking to achieve this goal with an intelligent master plan by London architect Norman Foster, which will include advanced building design, a clever HVAC (heating, ventilation, and air conditioning) system, CO_2-free traffic, and a large photovoltaic power plant outside the city.

Such projects will help architects and engineers create new plans for designing sustainable urban development. They may have little to do with reality because they are being planned as completely new projects and don't have a planning hierarchy relying on the behavior and decision-making involvement of the population. Time will tell whether these visions can be realized.

A Swiss Federal Institute of Technology Zurich (ETH Zurich) project in Ethiopia is working much more closely with local populations. The UN forecasts that the population of Ethiopia will grow from 80 to 125 million people in the next fifteen years, gradually turning villages into urban settlements. ETH researchers are now helping to develop construction technologies adapted to the country. Urban construction in Addis Ababa until now has followed models in Dubai or the United States, with high-rise buildings going up that consume an enormous amount of energy. Actually, the Ethiopian climate allows a building architecture in which there is no need for heating or cooling. The ETH Zürich project is reviving traditional building methods with indigenous materials like natural stone. Experiments are being done with techniques that require only a little cement in clay buildings. It is easy to forget that transporting building materials and producing cement during the building phase are major items in the CO_2 balance of a building calculated over its entire service life.

Even though there are many good examples of modern urban development, sustainability is difficult to achieve, especially in the poor city areas of emerging countries. From the outset, cities are subject to greater risk from climate change due to their high population densities and expensive infrastructures. This vulnerability increases when the outskirts of cities grow in an unregulated way and without proper infrastructure, when weak governments lose control, and when hopeless poverty spreads in slum areas. Then cities are even more at the mercy of climate change.

More than 3,000 cities around the world are built in coastal regions that lie less than ten meters above sea level. Nearly two-thirds of them are in developing countries in Latin America, the Caribbean, and Africa. Coastal countries like Bangladesh and the Pacific Island Nations, where 20 percent of the urban population lives on the seacoast, are especially endangered. With rising sea levels, the risk of coastal erosion increases. Computer simulations show that two million people on the coast of Egypt in the tourist area around Alexandria would have to leave their homes if the level of the Mediterranean were to rise just fifty centimeters.

There is still great uncertainty in climate research on how the frequency and intensity of major tropical hurricanes will change. But catastrophic storms in recent years give a foretaste of the potential outcome for cities. Every year, natural disasters affecting cities have been responsible for the death of thousands of people and have generated billions in damages. In 2005, around 80 percent of the destruction wrought by Hurricane Katrina in the Gulf of Mexico affected urban areas.

In wealthy cities it is taken for granted that people will protect themselves and take out insurance against extreme weather events, fire, and accidents. But urbanization often goes on in countries that can't keep up with rampant growth. Expensive office high-rises go up next to miserable shantytowns. Traffic breaks down daily, a thick

Is urbanization a solution? A New Yorker produces only one-third of the greenhouse gas emissions generated by the average U.S. citizen, and a resident of London causes about half the average emissions of a person elsewhere in Britain.

blanket of smog suffocates inhabitants, garbage piles up, and open sewers are breeding grounds for pathogens.

Nearly one billion people around the world live in slums under appalling living conditions. Prohibitive prices for land and housing force the poorest in the population to the outskirts of cities, to places where the risk of natural disasters is especially high. UN-HABITAT estimates that nearly half of the improvised dwellings in developing countries are threatened today by floods, mudslides, and storms, especially in slums and settlements without any official status. Today millions of people are migrating into cities, driven not only by hunger, but also by loss of their land and homes after a natural disaster or war—these migrants already include climate refugees. Overstrained metropolises are weakened further. Environmental protection, sewage systems, water provision, and city public health are then even more difficult to manage, and become very expensive.

A city is not just a dot on the map. The land surrounding it also belongs to its heart. Rising incomes in the cities of developing countries mean that the demand for agricultural products is increasing, and conversely, that farmers become dependent on the services of cities, especially for access to markets. This dependency is a very sensitive one. In 2009, Ketsana, a tropical storm, brought the worst flooding in more than forty years to the Philippines. More than 700,000 people lost everything they owned. The storm destroyed streets and bridges and more than 36,000 hectares of rice fields, along with plantations of high-quality grain. The destruction of rice and grain fields created a shortage in the supply of foodstuffs, and prices for food rose massively. A study in ten African states shows that it is often the urban population that suffers most when the food supply becomes scarce, even though urban incomes on average are higher than earnings in the countryside.

The best protection against natural disasters is offered by a good infrastructure with a functioning water supply system, sanitary installations, and a fully developed communication system. Even the poorest residential areas of a city would be helped if flimsy shacks were replaced with more stable housing, and if drainage systems were installed to carry off rainwater from these neighborhoods. The resettlement of poor neighborhoods to safer places could also help, although this makes sense only if the transportation system functions well enough to bring people to jobs and schools.

At the Earth Summit in Rio de Janeiro in 1992, Agenda 21 proposed that urban development should be surveyed regularly with regard to sustainability. A study published by the ETH Zurich points out that governments lack the political will—and the money—for sustainable development. Furthermore, relevant education and technology are also missing, so that public authorities in many small, fast-growing metropolises are often overwhelmed by their task. City governments are often unaware of which districts are endangered by floods or slides. Risk maps are an important instrument for taking precautionary measures before disaster strikes. Sometimes it would even be helpful to clear sand and garbage from drainage canals before the monsoon season began. But this kind of preventive action does require reliable weather forecasting services. The UN World Meteorological Organization (WMO) says that more than 60 percent of its member states do not have a functioning storm warning system—in the very countries that are most threatened.

Although a lack of governance makes cities vulnerable to the consequences of climate change, sustainable city planning can offer good opportunities. Firstly, it can protect against storms and floods, and secondly, sustainable urban planning can also make an important contribution toward mitigating climate change. City planning must be redefined to encompass sustainable principles and values. Cities, especially megacities, need not be gigantic machines wasting a lot of energy and emitting a lot of greenhouse gases. A single inhabitant of New York City generates only 30 percent of the emissions that the average American generates, and a single inhabitant of London is responsible for only 55 percent of average emissions per person in the UK. Large and densely built cities with good transportation systems and sustainable urban planning significantly help to reduce emissions. Taking measures in cities to mitigate climate change can lower the risks posed by natural disasters and at the same time improve living standards.

High density, short distances — cities offer ideal conditions for energy efficiency. Manhattan, New York.
James Leynse/Corbis

Turning a city like Cairo, the charismatic Egyptian capitol, into a sustainable and climate-friendly metropolis, is still a gargantuan task. *Amr Dalsh/Reuters*

Half the world's population lives in urban areas today, and that share is steadily rising. Eleven Asian cities have more than 10 million inhabitants; here Shenzhen in China. *Eugene Hoshiko/AP/Keystone*

At the mercy of storms and torrential rains—a lot of urban settlements in poorer countries like Brazil are built on unstable hillsides. Favela of Rocinha, Rio de Janeiro. *Janet Jarman/Corbis*

Demand for agricultural products in cities increases with income — food cultivation on the outskirts of the Chinese city of Guangzhou. *David Butow/Corbis*

To meet huge demand caused by growing populations, farmers will have to boost their production drastically in coming years, especially in Asia. Vegetable stand in Lucknow, Northern India. *Pawan Kumar/Reuters*

Urban agriculture reduces the need for transportation and could become an important factor in sustainable urban development in the future. Venezuela. *Giuseppe Bizzarri/FAO*

Mutual dependence – city dwellers need fresh produce from local farmers for whom access to urban customers is of vital importance—carts loaded with cabbages heading to a market in Cairo. *STR new/Reuters*

China plans to set new standards for sustainable urban living in its ecocity Dongtan, to be built on the island of Chongming. The project is still on the drawing board. *Artist's impression of Dongtan, designed by Arup*

THE SECOND "GREEN REVOLUTION"

Farmers are faced with a Herculean task. According to the UN Food and Agriculture Organization (FAO), they will have to increase food production by 70 percent by 2050, because in forty years more than nine billion people are expected to populate the planet. This task is all the more difficult because land and the supply of water have become scarcer and scarcer in many regions. Furthermore, climate change could aggravate the situation in the future, especially in developing countries where, as pointed out by the International Panel on Climate Change (IPCC), there is a high probability of more frequent droughts and floods. Water and land management adapted to new climate conditions is becoming a central factor for agriculture and food security.

Scientists at the Royal Society in the UK argue in favor of a sustainable intensification of agriculture. In simple words, this means increasing yields and intensifying farming without wasting water, salinating the soil, or overusing fertilizer and pesticides. This would also have the welcome effect that farmers would take precautions against drought and flooding, and reduce greenhouse gas emissions at the same time.

The second green revolution many experts are hoping for would have to meet quite different objectives from the first green revolution forty years ago. At that time it was just a matter of maximizing yields. Many believe that the green revolution of the sixties and seventies saved fast-growing populations in Asia from famine. New rice and wheat strains multiplied yields. Bangladesh tripled its rice production in forty years. From 1965 to 2008, the yield from basic foodstuffs like wheat, rice, corn, and soy increased around the world by an average of 1.5 percent per year.

Now and in the future, the challenge lies in increasing agricultural production using the highest ecological standards possible, and adapting practices to accommodate climate change. Unfortunately, the first green revolution put a mortgage on the land—today farmers use almost ten times as much nitrogen fertilizer as they did in 1960. This has made cultivation more expensive, and energy consumption for the production of fertilizers has gone up. Monocultures, more susceptible to changed climate conditions and diseases, have become common. Two-thirds of artificially irrigated areas were established after 1960. Today, 80 percent of the water used in Africa and Asia goes into agriculture; globally this figure is 70 percent. Rice alone consumes one-fifth of the world's use of water for grain. But yields of grain have been falling off in many places in recent years. The high-yield varieties planted in rice monocultures also contribute to reducing soil fertility.

There has also been a loss of fertile soils through erosion, and as a result, nutrients are leaching out—this loss is valued at $400 billion per year. About one million hectares of farmland are lost year after year because of salination caused by poor irrigation practices. Changes in sub-Saharan African states are horrendous. In some countries, the population has doubled in the last twenty years while the productivity of farmland has decreased by 40 percent. This can be partially blamed on overgrazing, which accelerates soil erosion. If global warming causes water to become even scarcer, then food production can fall even more.

Changing patterns in food consumption could further increase the consumption of water. With incomes rising in poorer countries, above all in China, eating habits are changing too. Meat consumption has tripled in the last thirty years and is expected to double again by 2050. Livestock relies on a plentiful supply of water and land. On average, 15,500 liters of water are needed to produce one kilogram of beef; this figure includes the cultivation of feed. One-third of global arable land is actually used to cultivate feed. The growing consumption of meat could slow down efforts to mitigate climate change. Livestock production alone, including the growing of feed and transportation, is responsible for 18 percent of all greenhouse gases emissions. The FAO has noted, however, that consumers are gradually shifting toward eating pork and poultry instead of beef. This trend would greatly reduce methane emissions since the digestive systems of cattle and sheep generate massive amounts of this gas, which has a very high greenhouse effect.

Agriculture is facing a radical shift. The only choices that remain are to expand areas of arable land or to increase the yield from hectares already under cultivation. The FAO says that additional fertile farmland can be made accessible only in sub-Saharan Africa and Latin America. But expanding areas for cultivation would be at the

DEVELOPMENT OF GLOBAL POPULATION
in billions

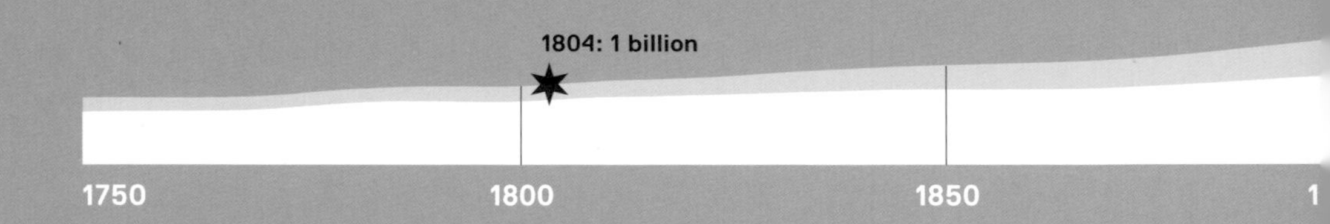

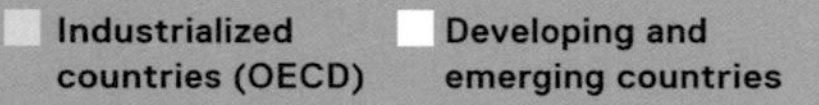

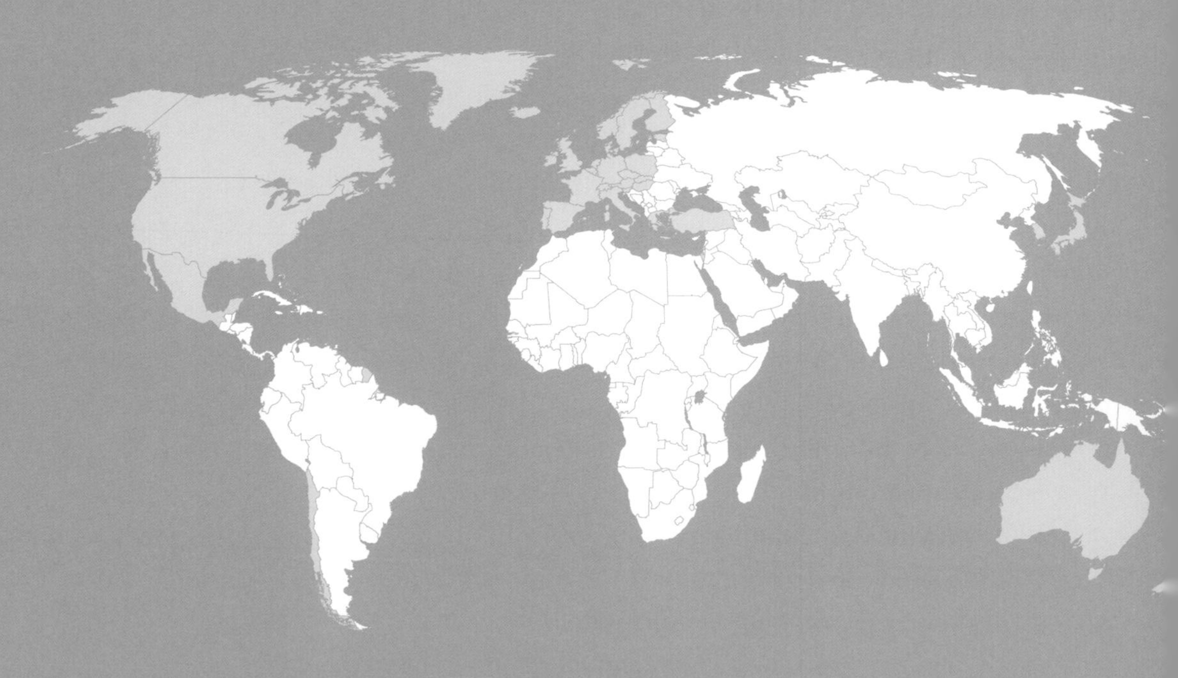

Source: UN Population Division/FAO/FAOSTAT in UNEP: The Environmental Food Crisis, 2009

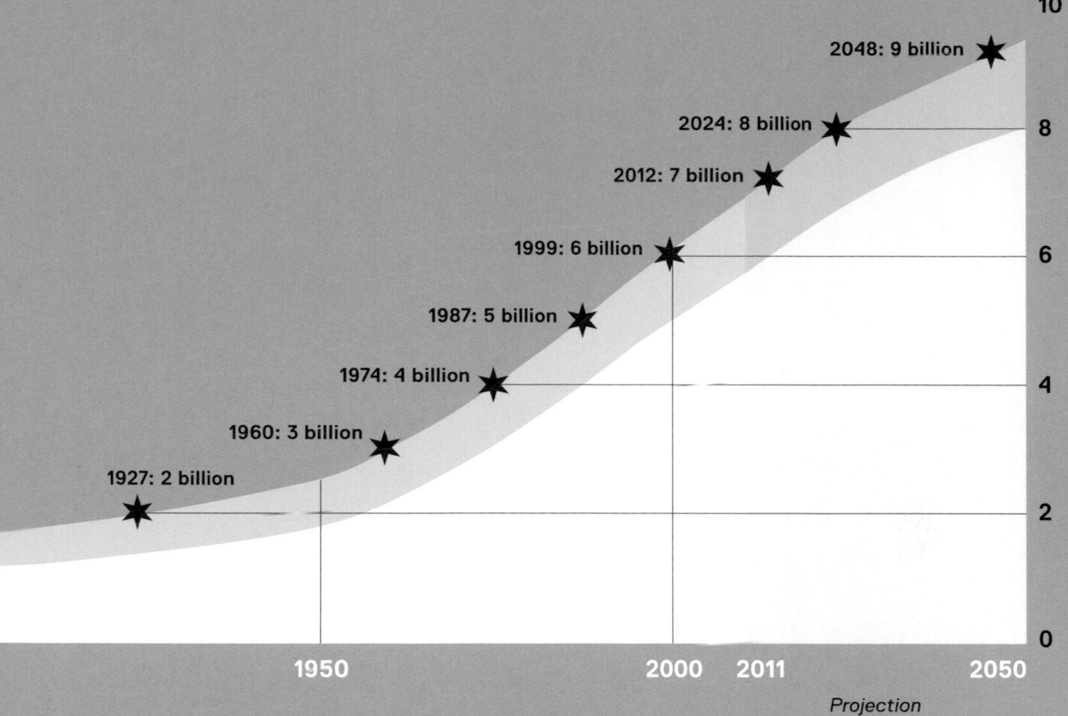

NUTRITION OF GLOBAL POPULATION
Kilocalories per person and day

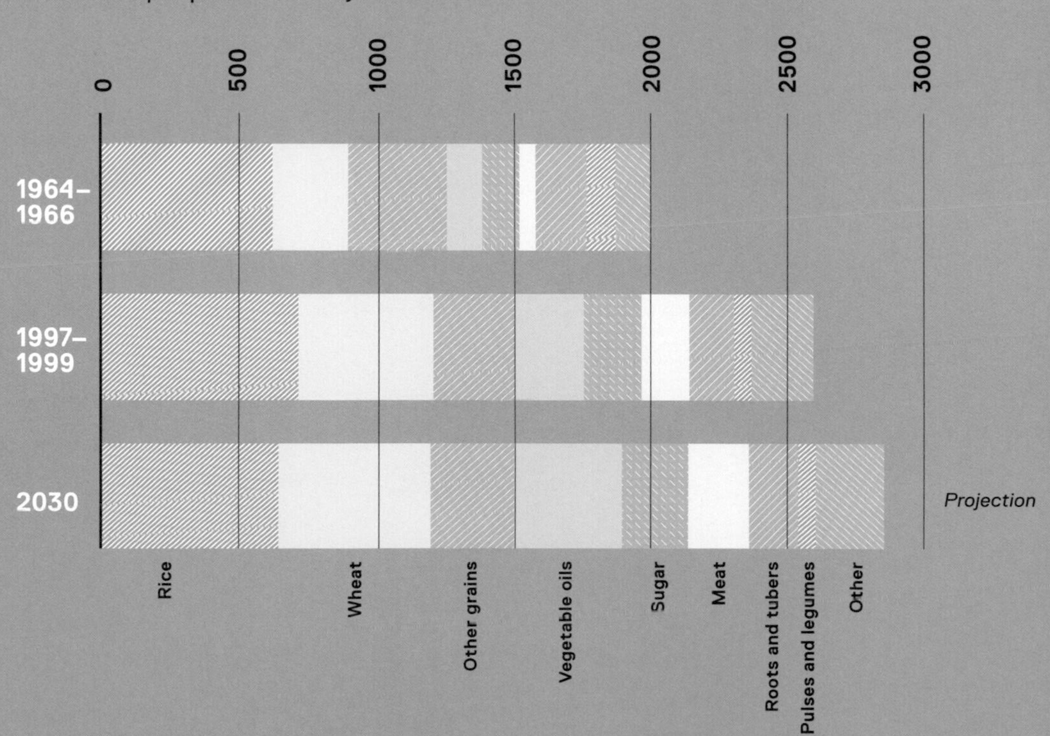

cost of rainforests, nature reserves, and ultimately the diversity of species and landscapes. The slashing and burning of rainforests today is responsible for about 17 percent of global greenhouse gas emissions →p. 518.

Sustainable agriculture offers a solution. A report from the Secretariat of the UN Convention to Combat Desertification notes that sustainable practices have the potential to make barely fertile land productive again. This is particularly significant because fertility is decreasing on a good 50 percent of cultivated land around the world due to erosion and salination of soil, or to a lack of water. The FAO says that around 8 percent of global farmland is already being sustainably cultivated.

Wherever farmers do not plow the soil, microorganisms produce less CO_2 through respiration. In undisturbed terrain, organisms in the soil and roots form a humus-rich soil structure with pores that store water. A soil covered with crop residues increases water retention and nourishes the soil, also protecting it from heat, wind, and water erosion and thereby cooling the ground and storing moisture in the soil.

Experience has shown that rotating crops of various types of grains and vegetables and planting tree cultures increase yield. Better soil fertility reduces the need for fertilizer, improving the energy balance of a farm. Returning crop residues to the soil instead of burning them also minimizes CO_2 emissions.

Many of the methods being proposed now are not new to farmers. For centuries they have had to adapt to varying environmental conditions, doing the best they can to preserve soil quality and retain as much water as possible. Farmers in developing countries use many sustainable cultivation methods, but their knowledge has remained largely undocumented. Some eighteen years ago, the Institute of Geography of the University of Bern began building up a digital database called WOCAT (World Overview of Conservation Approaches and Technologies), that holds records on years of long-standing indigenous experience—including all the socioeconomic, economic, and ecological benefits and drawbacks.

One example is rain-fed agriculture in southern Asia. During the monsoon season, 80 percent of rainwater from extreme rainfalls runs off the surface of the land. But this figure drops down to 10 percent if we take the entire season into consideration. The largest share of water is lost through evaporation. Farmers can cut water losses in half if they leave crop residues on the ground and do not plow the ground. Then they push seeds into the ground. Kenyan farmers developed this simple method of direct seeding some twenty years ago. Sometimes a coincidence or an inspiring personality can help a method become accepted. A respected farmer in the region around Mount Kenya gave this method a jump-start by encouraging small farmers to follow his example. Many of them couldn't imagine farming without plowing the ground, believing that crop residues would attract pests. In the meantime, the WOCAT database has grown and now holds information on more than 150 standardized cultivation methods that are being used successfully around the world.

Nevertheless, there are no patent remedies in agriculture. Cultivation methods must be compatible with the traditions of the country and region, the infrastructure, and socioeconomic factors. In Honduras, for example, farmers changed from traditional slash-and-burn agriculture to the Quesungual system. They applied mulch as a ground cover on corn and sorghum fields and planted trees. After three years, yields were significantly higher compared to yields from the earlier methods. Today they use the wood from the trees for firewood and wooden posts. Improved corn yields also mean that the volume of corn stalks and leaves has increased as well; this has become a profitable byproduct that farmers sell as cattle feed. Thanks to higher yields from plant cultivation, they can now lease out some of the land as pasture.

Another option for contributing to global food security under more difficult climate conditions is to combine agriculture with aquaculture. This approach is already widely used in Asia where profitable fish are raised in seawater basins. It would be feasible to fertilize cultures of salt-tolerant crops in dry coastal regions with aquaculture wastewater enriched with nutrients from fish excrements, or to cultivate sea algae, grasses, and mangroves, which in turn are useful as animal feed, food, or for the manufacturing of agrofuels. In the Negev Desert, for instance, tomatoes are irrigated with saltwater; in Ethiopia, glasswort, a halophyte, is watered the same way. Scientists in the United Arab Emirates are trying to grow glasswort in combined agriculture and aquaculture to produce biofuels. The advantage of this method is that very little fresh water is used. The disadvantage is that most fish in these aquacultures are fed with wild fish from the sea, which contributes to overfishing. It would be advantageous to invest in raising herbivorous fish in the future. Herbivore aquacultures would also help to mitigate climate change because fish would feed on plants, consuming far less energy and leaving less waste in comparison to processes for meat production. In a well-designed system, the nutrients from fish excrement would be recycled as fertilizer for fodder plants.

It is a controversial issue among scientists whether successfully increasing food production is possible when the climate changes and there is water scarcity without resorting to biotechnology. It does seem generally accepted that the cultivation of new genetically modified grains will be an option for food security in the future. The next plant generations will have to cope with various challenges: they must be resistant to insects, tolerant of herbicides, but also able to withstand heat, drought, and high salt concentrations in the soil. Recent estimates indicate that genetically modified plants are being cultivated today in twenty-five countries, among them fifteen developing countries. The first genetically modified plants, resistant to insects, were corn, cotton, rapeseed, and soybeans. These plants have increased productivity while decreasing the need for pesticides and herbicides and enabling no-till farming.

Even though there is no proof that the cultivation of genetically modified plants has negative consequences for the environment, or that genetically modified food is bad for consumers, public hostility toward genetically modified food is still substantial. Furthermore, small farmers in developing countries often can't afford to buy such seeds. Researchers in India are critical of the fact that 90 percent of cotton seeds are genetically modified (GM), and up to 95 percent of GM seeds are in the hands of the Monsanto group. Critics of genetic modification also fear that genetically modified plants could diminish the diversity of cultivated plant species, making plant cultures more susceptible to disease and more vulnerable to the consequences of climate change.

Whichever developments prevail, ultimately some 2.4 billion small farmers will decide whether a second green revolution is a success or not. Innovation and the ability to adapt to climate change depend upon governmental aid and upon the political and financial framework conditions created at the level of the United Nations and the World Trade Organization (WTO). The urgency of such measures was visible in 2008 when crop failures, caused by droughts and floods in major agricultural countries, substantially decimated global grain reserves. This could become the normal situation if sustainable agriculture is not pursued in the future. The consequence in 2008 was that prices for wheat, rice, and corn tripled. Exporting countries restricted the sale of basic foodstuffs, while importing countries bought higher amounts of grain to secure their food supplies. Speculators profited from the uncertain situation on the market. More than 100 million farmers and vendors in poor countries lost their income, and the crisis left about 40 million people hungry.

But agriculture in the Western world is not protected from the consequences of climate change either. The summer of 2003 in Europe had a mean temperature of about 3.5 degrees Celsius (6.3°F) above the average for the last century. The heat led to massive losses in yield for farmers, who harvested between 20 to 36 percent less grain and fruit. Although consumers barely noticed this on supermarket shelves, the heat wave showed how vulnerable agriculture is to extreme weather events. It seems unavoidable that heat waves will occur more frequently in the future because of global warming. However, there are also regions in the world where farming will benefit from warmer conditions. Various studies show that global agricultural productivity will not change greatly through climate change because the increase in food production in regions where conditions become favorable will compensate for losses in southern latitudes. How we deal with this situation is not a task for agronomists, but rather for politicians.

There are not enough wild fish to meet global demand. Aquaculture is an alternative, especially if fish can be fed on plankton, like the carp seen here in the Illinois River, USA. *Nerissa Michaels/Keystone*

The steady increase in worldwide demand for meat could undermine successes in mitigating climate change.
Slaughterhouse in Hamburg. *Christian Charisius/Reuters*

Pests breed rapidly in warmer climates—farm workers spray pesticides to control locusts in Tajikistan. *Nozim Kalandarov/Reuters*

Monocultures are energy-intensive; CO_2 emissions increase if produce is grown out of season. Strawberry field in France. *Martine Franck/Magnum Photos*

Mitigating climate change also means using less energy to manufacture and spread agrochemicals and fertilizers. Aerial spraying of pesticides over a rice field in Texas. *Jim Sugar/Corbis*

Regions expected to receive less rain will have to resort to artificial irrigation—orange groves in California's desert near Blythe, not far from the Arizona border. *Yann Arthus-Bertrand/Altitude*

Conventionally cultivated and genetically modified drought-resistant plant species for use in a changed climate — a potato gene bank at the International Potato Center in Lima, Peru. *Enrique Castro-Mendivil/Reuters*

Agricultural biodiversity lowers the risk of crop failures during periods of drought. A Mexican peasant woman drying heirloom corn varieties. *Philippe Psaila/Science Photo Library/Keystone*

Where marginal land is kept cultivated by local farmers, steep hillsides are less prone to erosion. Varzob Valley, Tajikistan. *Hanspeter Liniger*

Diversified terrace cultivation in Nepal helps to lower vulnerability to a changing climate — women carry dung to their rice and potato fields. *Hanspeter Liniger*

Traditional rice cultivation in Bali — terraces hold back water and lower flood damage even during heavy rains, allowing crops to prosper. *Hanspeter Liniger*

Covering the ground with a layer of organic material increases the soil's ability to absorb water and protects plants from heat. Farmers in Honduras. *Giuseppe Bizzarri/FAO*

BIOFUELS: A DOUBLE-EDGED OPTION

Biofuels are deemed an important factor in reducing CO_2 emissions. Because plants absorb and bind carbon dioxide from the atmosphere during cultivation, burning biofuels in car engines, heating systems, and power plants releases no additional greenhouse gas. Nevertheless, biofuels do not necessarily have a favorable overall environmental and economic balance.

Klaus Lanz

The German Renewable Energies Agency predicts that as early as 2020 renewable energy resources from domestic agriculture will supply 21.4 percent of Germany's car fuel, 13.1 percent of heating fuel, and 9.1 percent of the country's electricity. But for this to happen, the 1.6 million hectares of agricultural croplands currently producing energy crops such as rapeseed, grains, corn, and sugar beets would have to more than double to 3.7 million hectares. This is a good fifth of Germany's total agricultural land. An additional 800,000 hectares now lying fallow would have to be cultivated. Finally, farmers would have to stop growing food crops on over a million hectares and switch to energy crops instead. To compensate, they would have to considerably boost their food crop yields.

In principle, renewable fuels are carbon neutral, because the CO_2 released when fuels are burned is absorbed from the atmosphere during cultivation through the process of photosynthesis and thereby converted into biomass. The full climate balance is a bit more complex, as the amount of energy used in cultivating, harvesting, transporting, and processing a particular biofuel also enters the overall calculation. The more processing steps needed, the less favorable the balance is.

Today there are two main types of fuel produced from energy crops. The first is alcohol (bioethanol), produced by fermenting corn,

Gigantic corn silos in the Midwestern United States attest to the presence of an entirely new industry for biofuels. Ethanol facility in Windsor, Colorado. *Rick Wilkin/Reuters*

sugar cane, or grain, and suitable for technically adapted gasoline engines. Brazil uses it to meet half of its car fuel demand, and is also the world's leading ethanol exporter. Second, vegetable oils such as rapeseed oil or jatropha oil can be used directly as fuel or turned into a diesel substitute through chemical processing. The use of bio-kerosene in aircraft is still a long way off, but the aviation industry is working on producing it from raw materials such as algae, coconut, and jatropha oils.

The overall balance tends to be unfavorable, however, because so far only certain parts of cultivated plants, their oils, starches, and sugars, can be converted to fuel. Process technology to make use of grass, leaves, stalks, and other plant cellulose is not yet available. The large-scale conversion of cellulose, the main component of grass, straw, and wood, would be particularly important. Researchers so far have not been able to break down cellulose biochemically the way a cow's stomach does, to then convert it into biogas or ferment it to create alcohol. Not until these vegetable remains can be used to manufacture gasoline, diesel, or biogas will the CO_2 balance of renewable energy crops clearly improve. Biofuels produced from vegetable remains are now widely referred to as second-generation fuels.

Demand for palm oil is huge. The greater part of Malaysia's oil palm fruit is processed in industrialized countries by government-subsidized companies. Kuala Lumpur, Malaysia. *Zainal Abd Halim / Reuters*

The use of vegetable remains is however limited, because part of the harvested biomass always has to be returned to the fields where it degrades to humus, the net-like, carbon-rich structure that stabilizes soil and aids plant growth. If all plant remains were used to generate energy, humus in soil would diminish, making it less fertile and more prone to erosion by wind and rain.

Growing interest in renewable energy sources has led to the development of other biological technologies for generating energy. Algae belong to the third generation of biofuel sources. They can turn CO_2 in the atmosphere into biomass much more efficiently than plants growing on land. In theory, for the same amount of plant material, algae cultivation requires roughly five times less surface area than oil palm plantations, but it does need a lot of water. The idea is currently being tested for technical viability in pilot projects.

Energy planners are looking at other organic materials as well. Liquid manure, sewage sludge, wood residues and industrial wood, household compost, and catering waste are also suitable source materials for fuels. The Renewable Energies Agency reports that in Germany far more fuel was generated in 2008 from waste

Thirty years ago, Brazil pioneered the use of soy and sugar cane for energy; today, these energy crops meet half the country's fuel needs. Campos dos Goytacazes, Rio de Janeiro State, 2010. *Sergio Moraes/Reuters*

material than from energy crops—and there is still considerable potential to exploit.

A CHALLENGE FOR AGRICULTURE

A global boom in renewable fuels for transportation, heating, and the generation of electricity would have far-reaching consequences. Agricultural systems throughout the world would have to be completely restructured to meet this new challenge. Before governments and international organizations start to promote the cultivation of crops for fuel, they should carefully weigh the pros and cons.

To begin with, the impact of energy crops on the climate needs to be understood. If the intensive cultivation of energy crops involves the increased use of nitrate fertilizers and heavy agricultural machinery, additional nitrous oxide (N_2O), a greenhouse gas three hundred times more potent than CO_2, is generated in compacted soils. Biofuel produced in this way would in some cases have even more negative influence on climate change than oil.

To assess the future of biofuel, it is worth taking a closer look at Brazil, so far the only country with a full-scale commercial biofuel

Brazil is expanding agricultural lands for biofuels; displaced cattle ranchers resort to clearing Amazon rainforests to gain new pasture lands. *Paulo Whitaker/Reuters*

industry. Brazil introduced the cultivation and processing of energy crops in response to the oil crisis in the 1970s; the industry is now well established. The Dutch government in 2006 commissioned the universities of Utrecht and Campinas to carry out a detailed study of the Brazilian model — the overall conclusion was positive. The report said that Brazil had enough cropland to grow sugar cane both for ethanol and food products. Thanks to copious rainfall, sugar cane plantations did not need artificial irrigation and therefore did not put a strain on water resources. Even bagasse, the fibrous stalks of the sugar cane plant that could not be converted to ethanol, was useful and could be burned to generate electricity and heating.

Brazil is now planning to increase sugar cane crops eightfold and expand soybean plantations. These additional plantations will extend over many tens of thousands of hectares. While they will be located far from Brazil's tropical rainforests and pose no direct threat to them, 88 percent of the new biofuel acreage will claim land currently used for cattle grazing. In 2010, the magazine *Science* published a detailed model calculation and concluded that as a result, cattle ranchers displaced by sugar cane and soy would have no alternative but to clear new pasture lands in the country's rainforest regions.

Unparalleled boom—Indonesia's rainforests fall victim to huge oil palm plantations. Cultivating and using this kind of biofuel runs counter to mitigating climate change. *Beawiharta Beawiharta/Reuters*

The successful Brazilian model does however depend on certain conditions that are not found everywhere. Climate, soil, and water might not be as favorable in other places, and not everywhere are agricultural lands as extensive. When rainforests in places like Indonesia are burned down to make room for cultivating oil palm trees, both the CO_2 balance and the environmental impact are devastating. It takes hundreds of years for agricultural crops to absorb the amount of CO_2 released through forest clearing. This does not benefit the climate, and local ecology and biodiversity suffer severe damage →p. 518.

ENERGY VERSUS FOOD

The United Nations Food and Agriculture Organization (FAO) considers the cultivation of energy crops to be a very promising economic path for poor countries in Africa and Asia to follow. Many of these countries do not have oil or gas reserves, and are dependent on costly imports and fluctuating oil prices. African countries emerging from poverty have been set back for years because of abrupt increases in the price of oil in recent years. The FAO believes that these countries could become energy independent by cultivating energy crops and generating biofuel. This would also have the effect of providing farmers with a new source of income, gradually raising them out of poverty.

However, the obstacles to cultivating and processing energy crops are huge, especially for poor countries. Large investments and technical know-how are needed to process energy crops, regardless of whether they are fermented to ethanol or converted chemically to biodiesel. This could result in a new dependency on foreign technology, meaning that some of the added value would flow away from these countries. It is questionable whether they would genuinely reap the benefit of having affordable fuel at their disposal.

The farming of energy crops should not compete with the production of food or make use of forest regions and other valuable conservation areas. Since many countries do not have adequate cropland, the development of bioenergy would force them to increase their agricultural yields. This means using more fertilizers, pesticides, and heavy agricultural machinery, an approach to farming that almost inevitably leads to overfertilization, water contamination, and soil degradation.

Another option is the cultivation of fallow lands or areas with saline soil. Seed manufacturers propagate cultivating salt and drought-resistant plant varieties. But genetically modified and therefore expensive seed is hardly an option for poorer countries with meager foreign currency reserves.

In many cases it is difficult to confirm that the additional cultivation of energy crops does not have a negative impact on food supplies. When the global price of wheat skyrocketed in 2008, the World Bank saw a connection to the increasing production of biofuels. Even if many factors influence the global price of agricultural commodities, the link between agriculture and energy markets is clearly responsible for the tendency to turn land used for food production into acreage for energy crops. A case in point is Malaysia, which saw the demise of its newly developed biodiesel industry in 2008. High subsidies in industrialized nations artificially drove up the global price of palm oil, thereby making it totally unaffordable for local processing plants. Most of the available palm oil was instead bought up by the Western biodiesel industry, heavily subsidized by its various governments. Subsidies for energy crops in OECD countries amounted to roughly $15 billion in 2007.

Cultivating energy crops massively raises phosphorous and nitrogen fertilizer levels in the Mississippi, eutrophying and damaging fragile ecosystems in the delta and along the Gulf of Mexico coast. *NASA/Keystone*

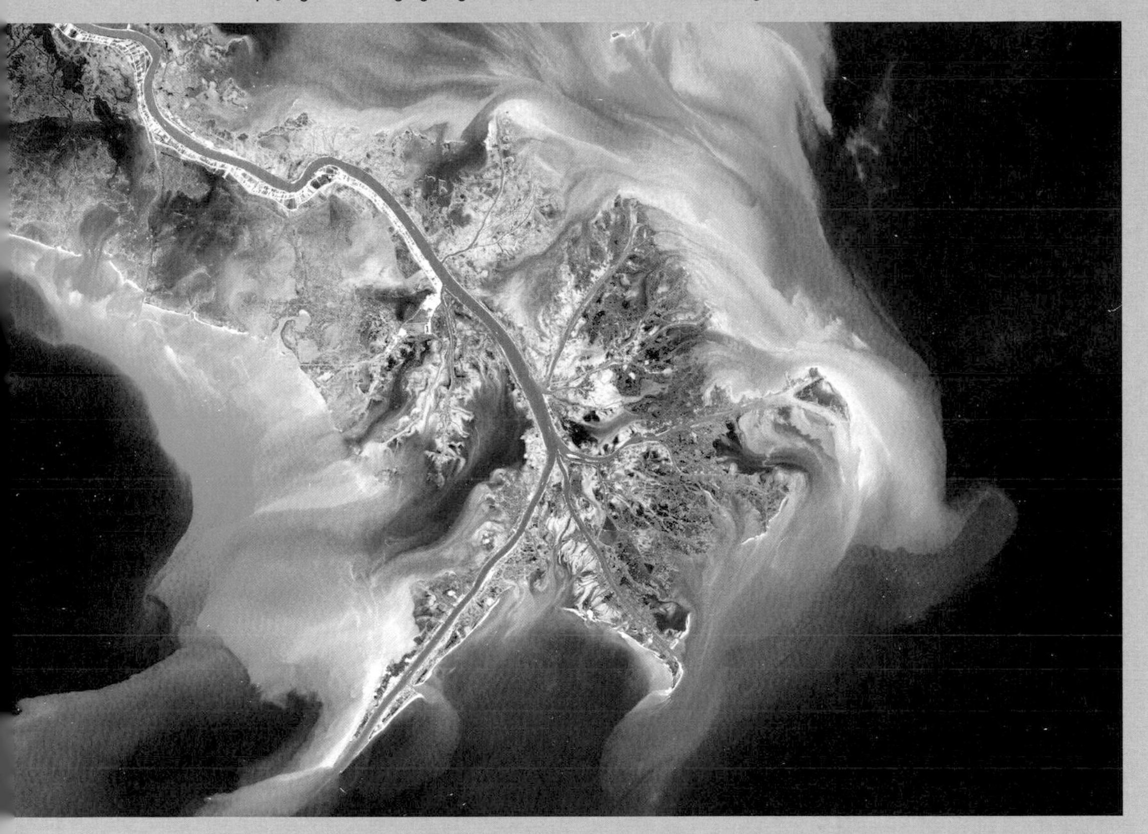

WATER SCARCITY SETS A LIMIT

In many regions of the world, water is even scarcer than cropland. In such regions, a shortage of water is already setting major limits on agriculture—even without taking energy crops into account. Every increase in agricultural production also requires more water, and additional demand is immense in view of some countries' biofuel plans. This means that regions with limited water resources cannot even begin to consider cultivating energy crops.

The International Water Management Institute (IWMI) has calculated that India will have to produce 60 percent more grain and more than double its sugar output by 2030 if it expects to feed its growing population. It will need at least 84 billion additional cubic meters of irrigation water to meet this challenge. If it also expects to cover 10 percent of gasoline demand by 2030 with ethanol, another 22 billion cubic meters of water will have to be made available. But India's water resources are already overused and groundwater levels, especially in Punjab, the country's breadbasket, are sinking dramatically.

The situation is no better in China. Rising demand for meat requires animal feed, which in turn needs enormous quantities of water—an increase of 45 percent is expected by 2030. China's water resources are already very scarce, especially in the northern part of the country, making it almost impossible to irrigate additional corn crops for producing ethanol. The IWMI is therefore urging: "It is high time discussions of biofuel energy put green energy into a blue context, and took water issues into account."

The United States government is also promoting biofuels, and passed a law in 2007 to increase ethanol production fivefold as a gasoline substitute by 2022. The plan is to produce most of the ethanol from corn, meaning that acreage would have to be expanded by 44 percent. The major corn-growing areas depend on artificial irrigation, and refining corn to ethanol also needs large quantities of water. Experts estimate that the bioethanol program will require at least an additional 5.5 billion cubic meters of water a year, but groundwater resources in many parts of the Corn Belt are already becoming scarce. To make matters worse, scientists say that the boom in corn cultivation is contributing to drastic environmental deterioration in the Gulf of Mexico. Increasing amounts of fertilizer are washed into the Mississippi by rain and carried to the mouth of the river where they feed algal blooms that decay, thereby depleting the oxygen in extensive zones where fish and crustaceans cannot survive.

OVERRATED POTENTIAL

Many unresolved issues about the future of energy crops dampen the hopes raised by the German Renewable Energies Agency. The idea of growing plants for energy basically makes sense, but it involves making enormous changes in global land use. Substantial subsidies for biofuels, particularly in industrialized nations, are accelerating this development. The extent to which governments intend biofuels to be used for global energy supply jeopardizes food production. For the first time ever, we are experiencing a direct link between food and energy markets, making food prices dependent on energy prices. Moreover, we need to establish international standards for the cultivation of energy crops in order to protect agricultural lands from overuse. Otherwise there is a high risk that a well-meant solution, intended to mitigate climate change, will endanger our food supply, biodiversity, and the conservation of protected nature preserves, forests, and wetlands.

MITIGATING CLIMATE CHANGE IS AFFORDABLE

"The American way of life is not negotiable." The American president George H.W. Bush spoke these words at the Earth Summit in Rio de Janeiro in 1992, and his son George W. Bush repeated them in essence ten years later when he withdrew the United States from negotiations on the Kyoto Protocol. The reduction of greenhouse gases laid out in the protocol would slow economic growth, said the U.S. government. With the same argument, Russia also hesitated for a long time to ratify the climate treaty.

The costs of mitigating climate change are crucial in determining international climate policy. Former chief economist for the World Bank, Nicholas Stern, clarified this concept initially in 2006 in his comprehensive report, the *Stern Review on the Economics of Climate Change.* He put forward the thesis that it would be far cheaper to prevent climate change than to pay for its consequential damage. Approximately 1 percent of the global gross domestic product (value of services and goods produced) would have to be spent each year to reduce greenhouse gas emissions enough by 2050 to prevent Earth from warming up by more than 2 to 3 degrees Celsius (3.6–5.4°F). But this was under the assumption that first steps would be taken immediately.

The Stern report for the first time attached a price tag to the mitigation of climate change. Some economists were astounded by this radical about-face because until then most had assumed that the cost of lowering greenhouse gas emissions would be far higher than the benefit of avoiding the damage caused by climate change. But more and more scientists consider such purely economic cost-benefit analyses unethical and cynical, because they require quantifying a human life in terms of dollars and euros.

The value of investments in mitigating climate change can be estimated only with great uncertainty today because their effect on the climate will probably not be felt until the second half of this century. The much more central issue is prevention: What is it worth to us, and what responsibilities are we willing to take on to protect the next generations? It raises the question of what risks economies will want to take in coming decades.

There is enough data available to draw a reliable picture. The World Bank says that nearly three million people lost their lives from 1960 to 2010 because of natural disasters; nearly one million people died in Africa alone during extreme droughts. Droughts and earthquakes generated around $2.3 trillion in damages around the world between 1970 and 2008. Poor countries were more strongly affected than wealthy countries. Although most of these disasters were not yet related to climate change, they nonetheless give us an idea of what to expect when disastrous events occur more frequently, as climate researchers say they will. From these figures, the World Bank concludes that preventive measures to mitigate climate change will pay off, especially because 1.5 billion more people will be living in cities forty years from now. The risk of damage from storms and flooding is especially high in urban centers because many people live together in restricted spaces and within an expensive infrastructure.

However high the cost of future damage might be, there are two reasons why converting the energy system is essential. If we don't take steps to lower greenhouse gas emissions, then our planet will warm up by more than 2 degrees Celsius (3.6°F), and humans and nature will have to pay a high price for the consequences. Moreover, reserves of coal, oil, and natural gas are limited.

Nevertheless, many economists are taking a wait-and-see attitude toward an immediate reorganization of the energy system. They believe it would probably be less expensive to invest in renewable energies on a large scale when they are mature and economically viable. But wind power and photovoltaics are good examples of new technologies maturing and costs rapidly decreasing when they are mass produced.

Now is the time to act. Power infrastructures in Europe and the United States have reached middle age, and in emerging and developing countries the enormous demand for energy is calling for a massive expansion of energy systems. Decisions made by the energy sector about which forms of energy it chooses to support in the future will be seminal. Against the backdrop of climate protection, the construction of new coal- and gas-fired power plants will lead to practical constraints. These power plants will have operating lives between thirty and sixty years, but it is within this period of time that greenhouse gas emissions are supposed to be reduced on a massive scale.

The energy sector and large-scale industry will move away from fossil fuels only if renewable energy technology is economically viable. Long-term international, national, and cross-national rules and incentives are needed to make this happen. The emissions market, in which emissions certificates are traded, plays a key role. A certificate gives the holder the right to emit one ton of CO_2. The market functions according to the cap-and-trade system. This means that there is a certain limit, the cap, on emissions rights owned by states and businesses. Within the framework of the Kyoto Protocol, the United Nations has allocated to industrial states a certain number of emissions rights that define specific CO_2 reduction targets. For example, the European Union has divided its Kyoto reduction target among its twenty-seven member states and prescribed correlative volumes of CO_2 that each individual state may emit (emissions rights). In turn, individual EU governments, in their own allocation plans, have promised emissions rights to individual sectors of the economy such as energy producers or cement manufacturers. Trading emissions rights within the European Union Emissions Trading System (EU ETS) has gone on since 2005 →p.476. Participation in the European emissions market is compulsory for EU businesses in relevant sectors, and more than 12,000 industrial plants in the areas of chemicals, steel, building materials, and energy production now participate in the scheme.

In general, the emissions market allows countries and businesses to choose the most cost-effective way to meet their reduction targets. As with any market, supply and demand determine prices in the cap-and-trade scheme. Depending on the market price for a ton of CO_2, it might be cheaper for a company to buy emission rights on the market than to reduce greenhouse gases itself. For other businesses, it comes out cheaper to invest in their own reduction measures. If a reduction target is exceeded, the surplus can be sold for a profit on the emissions market in the form of certificates.

Trading with emissions certificates has become a billion-dollar business. The EU ETS is the biggest trading system in the world, but there are also national CO_2 markets and exchanges. The European Climate Exchange, the largest trading platform, opened in 2008 in

London, and there are other CO_2 stock exchanges in Amsterdam, Oslo, Utrecht, and Paris. This has created new jobs as well for CO_2 traders, CO_2 financial specialists, CO_2 management specialists, and experts who develop new financial products in this sector. The global market value of the emissions trade in 2008 was around 100 billion euros, with the EU trading system, valued at around 70 billion euros, playing a dominant role.

Trading is done not only with certificates from industrialized countries—emerging and developing countries are also supposed to benefit from the system. Therefore, industrialized countries and businesses can trade with certificates they have purchased through climate protection projects in developing countries. This is the Clean Development Mechanism (CDM), which is based on the idea that emitters in wealthy countries may find it advantageous to invest in projects in developing countries because the same amount of money invested there could reduce more greenhouse gases than it would at home. Less developed countries benefit from the CDM providing them with modern and environmentally friendly technologies.

Thanks to investments from governments and businesses in the West, wind turbines of all sizes have been installed in developing countries, organic waste products are supplying biogas, and methane is being captured at waste disposal sites. The CDM Executive Board, tasked with monitoring the Clean Development Mechanism, has accredited approximately twenty-five different climate-friendly processes. These range from the production of renewable energy and energy-efficient plants to cement production with low climate impact, reforestation projects, and environmentally sound industrial processes. In this way, around 1 billion tons of greenhouse gas emissions are expected to be saved in developing countries by 2012. This is the same amount emitted every year by France, Spain, and Switzerland combined.

Nonetheless, experts and environmental organizations are calling for urgent reform to the CDM system. The projects are unevenly distributed, with a good 70 percent of CDM projects being carried out in India and China. They can also trigger counterproductive investments. An environmental organization, CDM Watch, warns of an increase in the production of the coolant HFC-22. The manufacturing process

“We are rich enough to mitigate climate change—and we are too poor not to do it.”

 Sigmar Gabriel, German politician

for this coolant, made primarily in China, leaves a byproduct called HFC-23, which is a greenhouse gas significantly more potent than carbon dioxide. The manufacturer can sell emissions certificates for destroying HFC-23. More than half of all CDM certificates come from the HFC-23 business, although there are only a few projects of this kind. The businesses involved are earning millions. They receive climate credits for the destruction of HFC-23 that are worth far more than the cost of destroying the gas. This warped incentive does not contribute at all to mitigating climate change.

Moreover, the CDM Executive Board is not efficient in investigating the quality of the projects proposed. Thousands of projects are in the pipeline, and it is questionable whether all will be registered by 2012. That is when the first period of the Kyoto Protocol runs out, and the states party to the climate convention have been negotiating for several years over whether there will be a second period at all →p. 500.

This political uncertainty is weakening the entire emissions market system. Indeed, trading will function only if binding reduction targets are set. If the EU does not resolve in coming years to reduce its greenhouse gas emissions by 30 percent, then demand for certificates will drop, and the CO_2 price will fall as well. However, only high prices will give the economy sufficient incentives to invest in climate projects and benefit from the emissions trade. The market—as experienced in 2008 and 2009—also reacts to economic crises. When companies produce less, they also emit less CO_2, and the demand for emission credits drops.

But even if the emissions market continues to be effective at the European level, then this instrument ultimately makes sense for global climate protection only when it is implemented around the world and when as many sectors of the economy as possible participate in it. China, Canada, Australia, and Japan are planning to set up emissions markets, although they are hesitant due to political and economic uncertainty. Skeptics like the Dutch economist Richard Tol therefore believe it more intelligent to levy CO_2 taxes on fuels than to bet on sharply fluctuating CO_2 prices that make investors feel uncertain.

The emissions market can motivate investors to invest in CO_2-free technologies. But a comprehensive conversion of the energy system cannot be achieved in this way, because the emissions trade doesn't cover important CO_2-emitting sectors such as transportation and households. In addition, the CO_2 price is too low to help capital-intensive technologies like CCS (carbon capture and storage) make a breakthrough. This technology is supposed to remove CO_2 from coal-fired power plant emissions and store it underground →p.303.

In addition to the cap-and-trade scheme, national and international regulations and subsidies will be needed until renewable energies can compete with fossil energy sources. There are several old and new ideas for fostering changes in the right direction. Photovoltaics in Germany and Spain are only marketable compared to coal and nuclear power because of feed-in tariffs paid for electricity fed into the grid. The automobile industries in Europe, Japan, and the United States have been making more efficient and cleaner vehicles since legal emissions standards have forced them to do so. In Britain there are plans to introduce a zero-CO_2 regulation for all new buildings. The installation of international emissions standards within various sectors can avoid distortions in competition and prevent businesses from moving away to countries with less stringent environmental laws.

At the 2010 UN Climate Change Conference in Cancún, the states party to the UN Framework Convention on Climate Change confirmed their intention to pay $100 billion a year into a climate fund, starting in 2020. These funds are meant for measures to mitigate climate change in developing countries, but it is uncertain where such vast resources will come from. Public and private sources come into question. An expert group gathered at the request of the United Nations, made up of finance ministers, experts in the finance community, and economists, believe that CO_2 taxes in industrial countries and in emissions markets are conceivable, as are CO_2 taxes on airline travel. It could also be feasible—although politically awkward—to introduce a step-by-step redistribution of those subsidies for fossil energy that are still flowing abundantly today in some industrial countries.

This cash flow is not likely to be enough in coming decades. The World Bank estimates that in developing countries alone, up to $175 billion a year will be needed to achieve the two-degree target. Another $75 to $100 billion will be needed so that the poorest countries can protect themselves from natural disasters. This is about 0.2 percent of the gross domestic product of all developing countries in the last ten years. Just for comparison—in reaction to the financial crisis, the global community mobilized trillions of dollars within weeks for ailing financial institutions to prevent economic collapse.

But efforts to reduce greenhouse gas emissions are effective only in the long run. Many countries in poor regions of the world are already suffering periodically from flooding and droughts—even without the influence of global warming. Each natural disaster sets back the affected country in its economic and societal development. In places where even the fight against poverty fails, there is no awareness, much less money, for effective climate protection.

As we can see from the example of Bangladesh, it is worth the effort to be as prepared as possible for natural disasters. Some 140 million people live in this country, whose area is roughly about three times that of Switzerland. Two-thirds of its area lies less than five meters above sea level. Every four or five years, these low-lying parts of the country are hit by devastating floods. In the last forty years, the government has invested more than $10 billion to protect city and country from inundation by constructing storm-proof buildings and erecting dams that prevent fertile farmland from being flooded with salty seawater. Although these measures are successful, annual damage from natural disasters still amounts to 0.5 to 1 percent of Bangladesh's gross domestic product.

An expert group of specialists from the United Nations, businesses, and nongovernmental organizations (NGOs), reviewing case studies from China, India, Mali, Britain, and the United States, has estimated that 40 to 70 percent of the losses expected by 2030 from natural disasters could be averted with timely countermeasures. The countries surveyed could potentially suffer damage costing up to 19 percent of their gross domestic products in the next twenty years. The experts identified more than 600 different measures for protecting the population from the consequences of climate change. These included changes in infrastructure, new technologies, behavioral adaptation, and, not least, insurance policies.

A World Bank project has been set up in Mali, for example, to protect small farmers from ruin due to crop failures; the project relies on weather-dependent insurance policies. The money a farmer borrows from a bank to buy seeds is partially or fully replaced through the insurance if weather conditions ruin a harvest. Such instruments are completely new for smallholders and have been difficult for them to trust. They have doubted the reliability of weather data, for example, which determines whether or not insurance will be paid out. Often farmers have not trusted weather records when an insurance payment was not granted. Nevertheless, insurance policies could play an essential role within the UN Climate Convention's adaptation strategy. The Swiss reinsurer Swiss Re has already insured 150,000 small farmers in Kenya, Mali, and Ethiopia as part of the Millennium Villages Initiative.

But in spite of adaptation measures, risks due to climate change will remain—and increase with the degree of global warming. If climate change cannot be mitigated, then it will become more and more difficult and expensive for individual countries to adapt. Measures to alleviate climate-related damages are not a substitute for reducing greenhouse gas emissions.

The highest rate of investment in climate protection can be expected in developing countries, because their economies will grow the most in the next twenty years. A UN Climate Framework Convention fact sheet explains that about half of global investments will go toward helping these countries catch up on development. But if this cash flow isn't successfully directed toward building renewable energy systems, then developing and emerging countries will miss out on sustainable development, and greenhouse gas emissions will continue to increase. If we don't make long-term political decisions and set ambitious targets for reducing emissions—whether at the international, national, or sector level—then there will be no way into a post-fossil world.

"It is not the pursuit of wealth that is the problem. The problem is that we define wealth as material possession."

 Dennis Meadows, economist, 2008

HUMAN BEHAVIOR: THE GREAT UNKNOWN

Energy efficiency is the magic word. In coming years and decades, steps to improve energy efficiency should save energy and reduce emissions of greenhouse gases. Around the world, the energy-saving fluorescent lightbulb is gradually replacing the incandescent lightbulb. Bulbs made of light-emitting diodes (LEDs), guaranteeing minimum consumption of electricity and an extremely long service life, are already on the market. Engines and cars consume far less fuel today and still perform as well as ever. More and more new electric appliances are rated in the highest category for efficiency.

Although engineers can calculate with precision how much energy is saved in this way, everyday life turns these calculations into theory. People behave in ways that thwart plans for saving energy. Researchers talk about a rebound effect. When more money is available for household budgets because of an energy-saving car (less money spent on gasoline), the household has more purchasing power. Japanese researchers discovered that a majority of the people asked to test new Toyota Prius hybrid cars actually drove significantly more kilometers per week than they did before in their old cars. After energy-saving lightbulbs were introduced in Great Britain, household lighting devoured even more energy, to everyone's astonishment. The savings effect is quickly lost if new and costly lighting is installed in a house or if living space is supposed to be ever more brightly illuminated. Experts in northern countries now assume that during the cold season more energy-saving bulbs will burn at home and in offices to compensate for lost warmth, because energy-saving bulbs give off less heat than traditional incandescent bulbs. All things considered, this could lead to higher electricity consumption.

Economists have often neglected calculating in the rebound effect. Years ago, critics were already finding fault with the famous Stern report →p. 506, which put the cost of global climate protection into numbers for the first time, but made practically no mention of the rebound effect. The same was true for the last climate report by the United Nations Intergovernmental Panel on Climate Change (IPCC).

And yet, the rebound effect is an old phenomenon. British economist William Jevons identified it as long ago as the mid-nineteenth century and warned that England's coal reserves would be used up quickly if steam engines became more efficient. His argument was that improved efficiency would lower the cost of coal per unit of energy, which in turn would stimulate the spread of coal-burning technology. This principle can be seen in the proliferation of computer technology. In early days, huge calculating machines called for a lot of personnel, capital, and time. Today's low-priced PCs save on working time and personnel. But are they environmentally friendly? The huge demand for PCs and servers and the round-the-clock operation of the Internet has led to enormous increases in power consumption. Furthermore, the manufacturing of computer chips is energy-intensive. The information society is one of those factors to blame for growing demand for electricity.

Nevertheless, the rebound effect remains a theoretical calculation for many economists. Indeed, there is a general lack of methodological instruments and relevant data to understand and quantify it. It becomes even more complicated when indirect effects need to be modeled. Where do people invest the money they save by using energy-efficient appliances and cars? Do they invest it in other energy-intensive goods or air travel? If yes, then there may even be a net increase in energy use, and experts talk of "backfiring."

Boosting efficiency can result in a lot of undesired effects if complementary measures are not taken at the same time. High demand for energy-efficient and inexpensive goods can have macroeconomic consequences that are harmful to the environment—production and the use of material and energy resources increase. If manufacturing is outsourced to cheaper countries, then transportation routes become longer. New factories are built and old ones are torn down, costing additional energy. If legislation designed to mitigate climate change becomes all too strict, then businesses may threaten to relocate to countries with less stringent environmental laws.

The message from the respected environmental economist Jeroen van den Bergh to governments, entrepreneurs, and individual consumers was: "Don't be fooled by good intentions!" Because most empirical studies map only a small section of the global economic system, the rebound effect is generally underestimated. We can expect it will be less strong in industrialized countries than in developing and emerging countries where people's needs for energy services and improvement aren't saturated yet.

But we are not helpless when confronted with the unpredictable behavior of people. The replacement of incandescent bulbs with energy-saving bulbs can be an effective energy policy if at the same time the price of electricity is raised. Many scientists are convinced that ambitious climate targets can be reached in many countries only if people give up things voluntarily as well. This means that individuals must redefine their lifestyles and become aware of the consequences of their actions. There are no studies of life in the slow lane. When people save time using modern technology, they have more free time for other things—which can cost more or less energy, depending on the activity.

Not to be forgotten are economic instruments like tradable energy or CO_2 credits that could counteract the rebound effect. But this assumes that the maximum allowable amount of energy consumption and emissions for businesses is defined by the state—and the market must bring itself into line with this policy. This concept would control the risk that, in spite of energy-efficient technologies, more energy is consumed in the end than expected because of the rebound effect.

A great deal can be done with modern technology to mitigate climate change, but with every innovation, important questions must be asked. How and where is it produced? Does it really save resources? Last but not least: How will people respond to it and how will it affect their behavior?

THE CARBON FOOTPRINT

Klaus Lanz

The carbon footprint represents the total amount of carbon dioxide (CO_2) emissions generated by an individual, and is closely related to a person's behavior as a consumer. People who buy a lot of things and tend to throw them away quickly, people who frequently use motorized means of transportation, who eat meat and frozen foods on a regular basis, or who live in poorly insulated houses, drive up their personal carbon balance.

However, consumers control barely half of their own carbon footprint. The remainder is at the discretion of manufacturers, who determine the energy balance of the goods they produce. When choosing an item to buy, a consumer usually cannot tell how much energy was involved in its production: mining the raw materials, manufacturing, transport, and packaging.

If global warming is to remain under 2 degrees Celsius (3.6°F) as agreed in Cancún, then each global citizen can produce no more than a maximum of 1 ton of CO_2 per year. The per capita value in Laos, Kenya, and Bangladesh is now 0.3 tons per year; it is less than 2 tons in emerging nations like Brazil and India, and in China it is 5 tons. The per capita value in wealthy industrial countries like Canada, Australia, and the United States is between 15 and 20 tons.

The individual CO_2 balance is determined not only by the amount of energy used, but also how that energy was generated. While a Swiss citizen has an average yearly footprint of about 6 tons, that of someone in Germany with the same living standard is nearly 10 tons. The difference lies in national electricity systems. Switzerland relies heavily on

carbon-neutral hydropower, while about 50 percent of electrical power in Germany is generated in coal-fired power plants.

A comprehensive switch to electricity to power all spheres of life, even indoor heating and cars, is sometimes promoted as the way to a 1-ton-CO_2 society. At second glance it quickly becomes apparent that electricity generated from renewable resources can hardly meet the enormous demand for energy of the Western lifestyle. Although the use of solar radiation, wind energy, hydropower, and geothermal energy can be extended significantly, there are still limits.

Future societies will have to generate carbon-neutral energy and use it in a much more intelligent and less wasteful manner. The yardstick will not be just the amount of CO_2 emitted, but also the amount of energy consumed, or to put it more precisely, the amount of electric power in watts a person consumes. For instance, the citizens of Zurich, Switzerland's financial capital, determined in a popular vote that their goal would be a 2,000-watt society. Average Swiss energy consumption is currently 5,000 to 6,000 watts per person. The 2,000-watt limit is therefore an ambitious goal, but it is realistic and can definitely be achieved if energy is used more conservatively and materials used more efficiently. A change in the consumer mindset would help people adapt to "using" things only when they are needed instead of "owning" them, as in car-sharing programs, for example.

The ecological footprint is an even more comprehensive concept used to assess individual consumption patterns. It calculates the effect of an individual's personal lifestyle by checking it against the area of natural space needed to produce clothing, food, and consumer goods, and handle the waste this has generated. The CO_2 emissions produced by individual lifestyles are compared to the area of forest required to reabsorb the CO_2. Global calculations using this approach estimate that humans currently consume the ecological capacity of 1.5 planets.

In principle, a specific amount of carbon dioxide emissions can be attributed to each consumer, thereby quantifying the concrete contribution each individual makes toward influencing climate change. This individual share, now widely referred to as a personal carbon footprint, depends on four basic factors: travel, housing, consumption of consumer goods, and eating habits.

People who travel frequently produce a lot of CO_2, although emissions depend markedly on the means of transportation. A car consuming 5 liters per 100 kilometers and driven 12,000 kilometers per year produces 2,370 kilograms of carbon dioxide; if it consumes 8 liters, emissions amount to 3,432 kilograms. Diesel vehicles produce around 6 percent more per kilometer due to diesel fuel's higher energy density. CO_2 emissions increase greatly with speed. Even a small, fuel-efficient car consumes at least 11 liters of gasoline per 100 kilometers when driven at 160 kilometers per hour.

The comparison to public transportation is instructive. Trains are three to five times more energy-efficient, with commuter trains consuming more energy because of their frequent stops. If we compare CO_2 emissions, the difference to cars is even greater since trains are powered almost exclusively by electricity. In Switzerland, where a high share of the electricity generated is carbon-free, the environmental impact of a train journey is roughly twenty-five times lower than that of a car going the same distance. Only if we compare a fully occupied small car with a half-empty train does the car win—although only as far as primary energy is concerned; the car's CO_2 emissions are still distinctly higher.

In countries like Germany, where trains run on electricity generated from a mix of energy sources, especially coal, or are fueled by diesel, the difference between cars and trains is not as marked. There, a small car consuming 4 liters per 100 kilometers in city traffic actually emits less CO_2 than trams or commuter trains. Once again, speed is a key factor. Not unlike cars that consume much more gasoline at high speeds, high-speed trains require double the amount of electricity at 240 kilometers per hour than they do at 160 kilometers per hour.

Air travel significantly affects personal carbon footprints. On a flight from Frankfurt to Singapore and back, each traveler is responsible for (depending on the emissions calculator used) between 2,400 (Qantas), 4,400 (myclimate), and 7,400 (atmosfair) kilograms of CO_2. These values vary because one of these calculators takes only pure CO_2 emissions into account, while another one includes an uplift factor to account for the particularly damaging effect of emissions at cruising altitudes of more than 9,000 meters.

The carbon footprint of frequent flyers shoots far above the average value. To improve it, we can either restrict our air travel—or else offset the carbon emissions we generate.

What is carbon offsetting? Basically, passengers compensate for the CO_2 emissions they induce by voluntarily supporting a project dedicated to reducing greenhouse gas emissions. These transactions are handled by various businesses and foundations that calculate the amount of CO_2 emissions produced during a specific flight and then initiate and invest in projects that mitigate CO_2. Most of these projects are located in poorer countries where reducing CO_2 emissions is less costly.

In practice, compensation programs can, for instance, distribute better and safer ovens that produce less CO_2 to people who otherwise do their cooking on primitive, smoky wood stoves. Indeed, this is even helpful for women, who then do not need to gather as much firewood as before and who are also spared from inhaling smoke and soot. It also reduces the amount of CO_2 released into the atmosphere. It benefits air passengers who no longer generate net additional CO_2 emissions because their contributions have made it possible for old and inefficient stoves to be replaced.

But even if damage to the atmosphere is avoided, the system still has its problems. In calculating CO_2 emissions, carbon offset providers end up with values that differ by as much as a factor of three. Some providers are less transparent than others about the projects they support. Moreover, it is questionable whether some of the reforestation programs offered will effectively absorb CO_2 from the atmosphere in the long run.

The success of the carbon offsetting system ultimately depends on the credibility of the businesses and foundations providing these services. They are carefully monitored and evaluated by environmental organizations and human rights organizations. It makes sense to use only the most reliable providers. The same applies to the choice of eligible projects. It is up to us to decide what we think of projects that convince farmers in India to use foot-operated pumps instead of diesel pumps to obtain drinking water, for instance, knowing that it takes a year's worth of pumping to compensate for a two-hour flight.

LOW-IMPACT HOUSEHOLDS

The two factors having the most impact in our homes are heating and electricity. Depending on the climate zone, at least 20 percent of a person's carbon footprint in an industrialized nation is due to heating because most households use oil or gas for this purpose. Electricity consumption has less impact on the CO_2 balance. Washers, dryers, and dishwashers are becoming more and more energy-efficient. However, increasing numbers of consumer electronic devices partly diminish these savings. Computers, telephones, and home entertainment equipment are responsible for a growing share of household electricity consumption, with standby appliances making a significant contribution.

Nowadays apartments are bigger, so private households have more lighting. Public spaces are also lit more brightly. Phasing out incandescent lightbulbs and replacing them with compact fluorescent lamps lowers consumption, and in a few years even more efficient light sources such as light diodes will be on the market. On the other hand, we can expect higher energy efficiency to encourage people to install more lights, which counteracts potential savings.

CONSUMER GOODS AND THE CARBON FOOTPRINT

At least a fourth of a person's carbon footprint is determined by the gray energy hidden in consumer goods—clothing, shoes, toys, entertainment electronics, and so forth. The global manufacturing of goods, which includes obtaining raw materials, production, transportation, and storage, is responsible for 36 percent of global CO_2 emissions. It is impossible to calculate the carbon footprint of most products because reliable information on their manufacturing processes is usually not available. In any case, a long life cycle is a particularly important consideration. Cheap, mass-produced, short-lived products have to be replaced frequently—involving renewed CO_2 emissions. Durable products that can be repaired are often more expensive to buy, but become more economical over time because they need not be replaced and have less impact on our environment.

An instructive example is the carbon footprint of an energy-efficient new car. Replacing an older car is generally regarded as beneficial to our climate. This is however only the case if the CO_2 emitted in manufacturing a new car, roughly 10,000 kilograms, is compensated for by lower fuel consumption. If a new car saves 3 liters of gasoline per 100 kilometers (equivalent to 7 kilograms of CO_2) compared to the car it is replacing, a positive effect on the climate will occur only after the new car has been on the road for more than 140,000 kilometers.

THE CARBON FOOTPRINT OF OUR FOOD CONSUMPTION

About 15 percent of our carbon footprint is determined by what we eat. The exact value depends a great deal on how much meat we consume, where the products come from, whether we choose to buy seasonal and organic foods, and finally on how much of our diet is made up of frozen foods. For every kilogram of meat or fish produced, 2.5 to 15 kilograms of CO_2 are emitted; for eggs and dairy products, the figures are 1 to 10 kilograms of CO_2; grains, potatoes, and baked goods are responsible for less than 1 kilogram; and fruit and vegetables grown in fields for only 0.15 kilograms per kilogram of produce. This is why a vegetarian diet lowers a person's carbon footprint by at least a fourth, and a vegan diet without eggs and dairy products by even more.

Seasonal produce has a much more favorable climate balance, as cultivating vegetables in greenhouses requires over ten times more energy than growing vegetables in fields. Distance does not affect the carbon footprint as much because transportation is only responsible for about 5 percent of emissions. However, products that are flown in from far away are especially energy-intensive, as are frozen foods, because their storage requires a great deal of electricity. Organic foods have a very good overall ecological balance, but their energy balance is merely a few percentage points better than that of conventionally grown produce.

LIMITED CONTROL

Consumers have only limited influence on a large share of their carbon footprint. There is not much information available on how much CO_2 is emitted in the production of consumer goods, food, gasoline, and building materials, or how large the carbon footprint of the parcel post and other services is. This part of a person's carbon footprint will continue to be determined exclusively by manufacturers. In many cases, the only decision left to consumers is whether to use disposable bags or carry purchases home in reusable totes.

Decades of cheap electricity and cheap oil have lead to a gigantic waste of energy in industrial nations, especially in the consumer goods sector. Transparency over gray energy in consumer goods could help people make an educated choice and decide to buy products manufactured in an energy-saving way, bringing about a gradual change in production processes. However, a political decision to raise energy prices would be far more effective, since this would trigger a rapid change in our industrial system and significantly contribute to mitigating climate change.

The Earth at night—brightness reflects prosperity. The lights visible offshore are gas and oil platforms in the North Sea and fishing fleets in the Pacific. *Planetary Visions/Science Photo Library/Keystone*

Celebrating Christmas with 25,657 light bulbs and 7,135 watts. Bremen, Northern Germany.
Jörg Sarbach/Keystone

Garbage dumps bear witness to excessive consumerism and the short life cycles of our consumer goods.
Martin Rütschi/Keystone

The electronics industry continually develops new technologies and products. These innovations afford us more comfort but may require more energy. *Martin Katz/Xinhua Press/Corbis*

Air conditioners and heaters cause electricity consumption to soar when construction is not adapted to the local climate. Typical high-rise apartment building in Sham Shui Po, Hong Kong. *Luke Chan/Getty Images*

Sidewalk café in Barcelona in February of 2011—the temperature outdoors is 15 degrees Celsius (59°F).
Stanley Wong

In love with cars—automobile fans watch a public screening of a Formula One race in Germany, 2010.
Martin Meissner/Keystone

Amusement park in the United Kingdom, 1999. *Peter Marlow/Magnum Photos*

Sunday outing, United Kingdom, 1994. *Martin Parr/Magnum Photos*

 Stretch limousine in Russia, 2009. *Dagmar Schwelle/LAIF/Keystone*

A flight from Europe to Singapore and back has about the same impact on our climate as the total average CO_2 emissions produced by an inhabitant of Europe in an entire year. *Ragonese-Scardino/Keystone*

 Electricity consumption on a cruise ship is the same as that of a small town. *Mauro Fermariello*/Keystone

 Piazza San Marco, Venice. *Paul Seheult/Corbis*

UCCI
GUCCI

Bargain hunting—no time to think about a product's energy balance. Tolleson, Arizona, USA.
Michael Ging/Keystone

TARGET

 California, USA, 1967. *Robert Landau/Corbis*

Market Basket
TENDERAY
BRAND®
BEEF
DOOR

GLOBAL CLIMATE POLITIC

BETWEE
POWER A
PARALYS

Heidi Blattmann

S

N

AND

IS

The atmosphere knows no national boundaries, nor do the weather conditions that are influenced by changes in the atmosphere. Therefore the battle to mitigate climate change can be successful only if states decide to act together. It is the global nature of this problem that confronts politics with very special tasks, and it is global not only because it affects all continents and oceans, and all countries and populations, but also because it encompasses nearly all areas of life. The effort to set guidelines for climate-friendly development in legally binding agreements is behind the international negotiations taking place under the auspices of the United Nations, and especially behind the annual conferences of the UN Framework Convention on Climate Change (UNFCCC), which receive so much media attention every year.

The immense complexity of the task starts with climate science, and the range of issues affected is enormous. Climate change spans all realms of time, from many millions of years of prehistory to scenarios for the next hundred years, from the split second of a photochemical reaction to geological processes evolving over millennia. The spatial variability of climate-related factors is no less enormous, encompassing the dimensions of molecules and aerosols in the nanometer range as much as the movement of continents and oceans measured in thousands of kilometers. It involves geological, physical, and chemical processes as well as the entire biosphere and all its diverse interactions → p. 133.

Mapping the vast breadth and depth of these processes in their varying dimensions poses major problems to climate modelers. Even if all correlations could be identified in detail and accurately represented in simulation models, the most powerful computers in the world would be unable to cope with the task for a long time. Climate scientists have to make do again and again with gaps in knowledge and simplifications, so-called parametrizations. This in turn leaves room for debate and differing interpretations, and makes it impossible to cast predictions that are beyond a shadow of a doubt—a problem which more or less characterizes every scientific forecast.

But it is not just the climate that is complex. Our society, the economic system, and politics are determined by a multitude of factors and their interaction is understood only to a certain extent, making it difficult to predict their development. An attempt to steer this system would be a challenge even if all people, communities, and countries, with all their different options for action, had exactly the same interests and absolutely agreed on all means, objectives, and priorities. Even then, not everyone would react in the same way to the same situation. The tremendous acceleration of data processing and the dissemination of information could make it even more difficult to manage with any degree of precision the globally interconnected economic and sociopolitical system and point it in the right direction.

Climate policy is not only about influencing countless control loops in nature, the economic system, and society in the light of looming climate change as diagnosed by climate scientists. It is above all about basic ethical values that encompass the issue of justice over generations. At the heart of the problem is carbon dioxide, the most important greenhouse gas generated by human activity, which has become a symbol for the abundant use of energy, and represents wealth and prosperity, comfort and power. Negotiations are about fossil fuels, the lifeblood of today's industrial world. Different stakeholders have different interests that are at least as pronounced on the world stage as in individual countries and communities with their own political systems. The cement manufacturer thinks differently than the forester; Europeans have different cares and concerns than do Indians, Arabs, or Americans; the wealthy find other things important than do recipients of welfare payments; the young focus on other issues than the elderly; and the powerful and rich have options available to them that the powerless and poor can only dream of.

Finding solutions together, in consensus with all countries, is the ambitious goal of international climate negotiations—this is the prerequisite within the framework of the United Nations. Before the climate conference in Copenhagen began, Tony Blair, former British prime minister, said that climate negotiations were more difficult than any other negotiations he had ever had experience with, Northern Ireland and the Middle East included. A famous metaphor by German sociologist Max Weber (1864–1920), who once wrote that politics is a strong and slow boring of hard boards that takes both passion and perspective, could not more appropriately reflect this sentiment.[1]

But not only the drilling of hard boards in the international political arena will determine the success of efforts to reduce greenhouse gases—as important as an internationally binding agreement is. An equally important role will be played by the general political development of the world, especially in the energy sector. And it will be even more important to see how individuals, families, village communities, towns, metropolises, and individual countries, as well as businesses and whole industry sectors respond to the findings of climate science. Beyond legally binding regulations, there are already a vast number of activities in place to protect the climate. They call for a high degree of solidarity and foresight. But these grassroots initiatives together set a trend in motion that not only reduces emissions but also motivates others to do the same, and at the same time may just have an impact on the international negotiations, whose slow progress is frustrating for so many.

1 Max Weber, *Politics as a Vocation* (Minneapolis, 1965); originally published in German as *Politik als Beruf* in 1919.

ENIAC, the first electronic computer of the US Army, was also used after 1950 for numerical weather forecasting; it was programmed by plugging cables. *Bettmann/Corbis*

The Weather Ball, a Massachusetts Institute of Technology lab for studying weather patterns, on top of Mount Washington in New Hampshire, 1960. *Orlando/Three Lions/Getty Images*

BROWN UNIVERSITY *Providence, Rhode Island • 02912*

DEPARTMENT OF GEOLOGICAL SCIENCES

(401) 863-2240

December 3, 1972

The President
The White House
Washington, D. C.

Dear Mr. President:

Aware of your deep concern with the future of the world, we feel obliged to inform you on the results of the scientific conference held here recently. The conference dealt with the past and future changes of climate and was attended by 42 top American and European investigators. We enclose the summary report published in Science and further publications are forthcoming in Quaternary Research.

The main conclusion of the meeting was that a global deterioration of climate, by order of magnitude larger than any hitherto experienced by civilized mankind, is a very real possibility and indeed may be due very soon. The cooling has natural cause and falls within the rank of processes which produced the last ice age. This is a surprising result based largely on recent studies of deep sea sediments.

Existing data still do not allow forecast of the precise timing of the predicted development, nor the assessment of the man's interference with the natural trends. It could not be excluded however that the cooling now under way in the Northern Hemisphere is the start of the expected shift. The present rate of the cooling seems fast enough to bring glacial temperatures in about a century, if continuing at the present pace.

The practical consequences which might be brought by such developments to existing social institutions are among others:

1) Substantially lowered food production due to the shorter growing seasons and changed rain distribution in the main grain producing belts of the world, with Eastern Europe and Central Asia to be first affected.

2) Increased frequency and amplitude of extreme weather anomalies such as those bringing floods, snowstorms, killing frosts etc.

Geologists warned President Nixon late in 1972 that a research conference at Brown University had come to the conclusion that global cooling was a real possibility. *Facsimile of original letter/Daphne Gemmill*

An iconic view of our vulnerable planet—astronaut William Anders photographed the rising Earth in 1968 during the Apollo 8 mission, the first manned spaceflight around the moon. *NASA/Keystone*

A Sunday walk in 1973 on West Germany's autobahn, declared car-free for the day—the OPEC oil embargo made the degree of dependence on fossil fuels strikingly clear. *Moesch/dpa/Keystone*

The environmental movement began to take shape in the 1970s—a 1983 demonstration against forest dieback in Geneva. *STR/Keystone*

After a Europe-wide search, barrels of dioxin from Seveso were finally found in Anguilcourt-le-Sart near St. Quentin in 1983; their contents were later burned at Hoffmann-La Roche in Basel. *Roche*

After the 1986 Chernobyl nuclear disaster, the area around the Ukrainian plant became an exclusion zone; fallout from the explosion increased radiation levels in much of Europe. *Igor Kostin/Corbis*

The Berlin Wall fell on November 9, 1989; tens of thousands of people poured into West Berlin the next day from East Germany to celebrate their new freedom. *EPA/Lehtikuva/Keystone*

In the 1970s, CFCs came under suspicion for destroying the ozone layer; their use in spray cans was restricted or banned in many countries. *Peter Turnley/Corbis*

The influence of environmental NGOs grows—a Greenpeace campaign in 1995 prevents Shell from dumping the Brent Spar oil platform into the Atlantic. It was later dismantled on the coast. *Hugo Bergsaker/Keystone*

IN THE WAKE OF OZONE PROTECTION

Early discussions about possible climate change and its political implications date back decades. But real momentum on the issue was gained only at the beginning of the 1990s, in the wake of efforts to protect the ozone layer. The problems waiting to be solved in climate politics are far more complex however than those underlying the depletion of stratospheric ozone.

The first attempt to define a global climate policy in specific terms was made in the mid-1980s. The World Meteorological Organization (WMO), the United Nations Environment Programme (UNEP), and the International Council for Science (ICSU) held a conference in October 1985 in the Austrian city of Villach. The final declaration issued by its nearly ninety participants—scientists, administrators, and representatives of international organizations—left hardly any doubt that greenhouse gas emissions generated by human activity would lead to global warming. The declaration challenged governments to incorporate the results of the Villach conference into their planning, to better inform the public, and, finally, to gather scientists and policy makers together to work actively on clarifying the effectiveness of possible measures and policies. Governments were to make more money available for research, and participating UN organizations were to regularly prepare reports assessing the level of knowledge and the practical implications of climate change. The declaration also recommended that a global convention should be considered, if necessary.[1]

The danger of global warming, however, had been discussed long before 1985. Early in the twentieth century, Swedish Nobel Prize winner Svante Arrhenius, referring to the work of earlier scientists like Jean Baptiste Joseph Fourier and John Tyndall, formulated the well-known theorem of global warming, which holds that a doubling of the carbon dioxide concentration in the atmosphere will lead to 4 degrees Celsius (7.2°F) of warming. Shortly after World War II, the possibility of an undesirable change in climate began to seriously concern public servants and politicians as well as scientists. In the United States, major computer centers began running the first numerical climate models. By 1965, the President's Science Advisory Committee noted in a report on environmental quality that human activity influenced the climate, with consequences affecting the entire globe. In the early seventies, when environmental problems were receiving more and more public attention, studies commissioned by the Massachusetts Institute of Technology (MIT) looked at the climatic effect of rising CO_2 concentrations and warned of a possible global disaster. Other causes for concern included large-scale plans by the Soviet Union to divert Arctic rivers toward the south for irrigation purposes or to close the Bering Strait with a gigantic dam to make permafrost areas fertile.

But in those years it was not the fear of global warming, but rather of a new ice age that turned the climate of the future into a widely discussed issue, especially in the United States. Paleoclimatologists had started to realize that we are living in a transition period between ice ages, and that it could soon come to an end. In 1972, a group of scientists therefore wrote a letter to President Nixon in which they warned of the "very real possibility" of imminent global cooling having a dangerous effect on food production—not least pointing out that the Soviet Union might already be incorporating this development into its international policy. At the same time, other scientists called upon the White House to approve a comprehensive program for climate research. Unusually cold winters like the one in 1972–73 further nurtured these fears. Consequently, the Nixon Administration took a first step in 1974 and created an interagency working group to deal with the issue. In Europe as well, scientists began to refer to the dangers of climate change, but they were more concerned about global warming as a consequence of increased carbon dioxide concentrations.

At about the same time, Mario Molina and Sherwood Rowland in the United States gave warning of another pending global disaster. In a groundbreaking article published in the scientific journal *Nature* in 1974, they put forward the theory that the release of synthetically manufactured chlorofluorocarbons (CFCs) into the atmosphere would in the long run destroy the stratospheric ozone layer. This research later brought them, together with Paul Crutzen from the Max Planck Institute for Chemistry in Mainz, the Nobel Prize for Chemistry in 1995. Crutzen contributed significantly to the understanding of the atmospheric chemistry of nitrogen. Extremely long-lived CFCs were being used in enormous quantities, especially in spray cans, but also as refrigerants and to manufacture foam. In 1971, just three years before this alarm was sounded, American chemist Harold Johnston had already drawn attention to the fact that the stratospheric ozone layer was potentially threatened by a plan to build a fleet of supersonic transporters (SSTs) that would emit nitrogen oxides at high altitudes, possibly leading to a breakdown of the ozone layer. Both of these threatening scenarios were discussed intensively in the U.S. Congress.

Scientists explained that a breakdown of the stratospheric ozone layer would be fatal for life on earth. The ozone layer filters out shortwave and therefore especially energy-rich ultraviolet rays of the sun, preventing them from reaching the earth's surface where they can inflict severe radiation damage to human beings, animals, and plants. Plans to build the supersonic fleet were abandoned, not least for economic reasons, and the warning issued by Molina und Rowland soon led to concrete measures being taken to reduce CFCs. Using measurements taken in balloons, in 1975 American scientists were able to prove for the first time that CFCs were reaching the stratosphere. As a consequence, in 1978 the United States, and other countries soon afterward, banned the use of CFCs in spray cans.

Only seven years later, the Vienna Convention for the Protection of the Ozone Layer was signed by more than two dozen countries. As early as 1987, a protocol within the framework of the convention, aiming to reduce CFCs, was drafted and came into force less than two years later as the Montreal Protocol on Substances that Deplete the Ozone Layer. In 1990 and 1992, amendments to the protocol followed that really gave it some teeth.

These successes were enabled not least by the political pressure arising in 1985 from the discovery of a hole in the ozone layer opening up over Antarctica in September and October. Admittedly there was no scientific explanation for this phenomenon yet. But the hole made the urgency of the matter obvious. Furthermore, rumors were beginning to circulate about a decrease in the ozone layer over the Northern Hemisphere; this was officially confirmed three years later by the Ozone Trends Panel of the U.S. National Aeronautics and Space Administration (NASA).

As a result, the American chemical corporation DuPont, until then a forceful opponent of regulatory efforts and the largest producer of CFCs in the world, announced that it wanted to limit applications for CFCs and develop substitute materials for many areas of use. This political about-face gave rise to quite a bit of speculation about why DuPont changed its mind. One argument was that the corporation had publicly declared early on that it would stop manufacturing CFCs if there were trustworthy data showing that these compounds posed a hazard to health. A second line of reasoning was that fears of facing lawsuits from skin cancer patients who might trace their suffering back to a depleted ozone layer may have played an important role. In any case, American negotiators, supported by DuPont's decision, were able to announce the decisive breakthrough in negotiations in 1986 in Geneva, and now even promoted the adoption of the protocol.

The 2010 report from WMO and UNEP indicates that stratospheric ozone is no longer depleting and may return to pre-1980 values before the middle of the twenty-first century. The British Antarctic Survey expects that by 2080 the annual thinning out of ozone over the Antarctic will no longer be any more severe than in the 1950s. The Montreal Protocol had the additional effect of being very beneficial for climate protection because many substances that break down ozone are also extremely effective greenhouse gases. This ozone assessment report, published in spring 2011, concludes that a side effect of the Montreal Protocol has been to lower greenhouse gas emissions in 2010 by the equivalent of 10 billion tons of CO_2. This is about one-third of the total global carbon dioxide emissions caused by the burning of fossil fuels and the production of cement. In other words, this is almost twenty times more than the yearly reduction of emissions that the Kyoto Protocol is supposed to achieve. The gases already regulated under the Montreal Protocol are in fact explicitly excluded from the UN Framework Convention on Climate Change and the Kyoto Protocol and are not taken into account for reaching the Kyoto targets. Nonetheless, the numbers show how greenhouse gases in the atmosphere would have increased much more without the Montreal Protocol.

The success story of efforts to save the ozone layer is referred to again and again as a model for climate negotiations. Facing public pressure, and with the support of industry, the world community of states had put a global agreement into effect within a relatively short period of time under the auspices of the United Nations, even though many scientific questions had been left answered. Why wouldn't it be possible to do the same for the climate? There were open questions here as well, and the conviction was growing that global warming posed an immediate threat and that action must be taken. The efforts made to protect the ozone layer therefore considerably influenced attitudes toward climate policy.

However, there were and still are crucial differences between policies for protecting the ozone layer and policies for protecting the climate. In the context of the Montreal Protocol, it was a matter of finding regulations for a manageable number of substances that were generally easy to replace. Only a few major chemical corporations and business sectors were relevant for the problem of ozone depletion, certainly not comparable to the number of economic sectors and enterprises that play a role in climate change. This was why DuPont's decision to give up resisting international regulation was so significant for the ozone protocol. In contrast, there is no comparable powerful business or institution that by itself could influence the course of action for mitigating climate change.

Nonetheless, success in regulating CFCs strengthened the determination of Mustafa Tolba, then director of UNEP, to make use of that environmental momentum to protect the climate as well. UNEP had made efforts since its founding in 1972, as had WMO since the 1960s, to research and document climate change. WMO held the first World Climate Conference in Geneva in 1979. Now Tolba aspired to organize a global climate convention. Just six months after the Vienna Convention for the Protection of the Ozone Layer was adopted in 1985, WMO, UNEP, and ICSU invited parties to the aforementioned meeting in Villach to set up the groundwork for negotiations. Tolba sent the conference's final declaration to U.S. Secretary of State George Shultz and called upon him to become politically active and to initiate negotiations for the drafting of a climate convention.

The Reagan Administration first proposed setting up an intergovernmental mechanism to prepare an international state-of-the-science report; this mechanism later finally became the Intergovernmental

Panel on Climate Change (IPCC). Then things happened in rapid succession. Late in 1988, the UN General Assembly gave UNEP and WMO the green light for establishing the IPCC and called upon both organizations to make sure the panel submitted a comprehensive report as quickly as possible. One year later, the Assembly also supported the UNEP proposal to begin preparing a draft for a climate convention. This was to include the IPCC findings as well as those of the second World Climate Conference scheduled for the fall of 1990. In its final declaration, the World Climate Conference in turn supported the rapid drafting of a climate convention to be ready for signing at the big United Nations Conference on Environment and Development (UNCED) to be held in Rio de Janeiro in June 1992, later known as the Earth Summit. Late in 1990, the UN General Assembly mandated the start of negotiations. Here as well, the objective was to prepare a convention text ready for signing in Rio.

With the drafting of a UN Framework Convention on Climate Change (UNFCCC) and with the IPCC serving as an advisory scientific committee, the foundation was laid for defining the international climate policy in coming decades, similar to the work on the Vienna convention that had laid the ground for the success of the Montreal Protocol. However, political negotiations have proven to be significantly more difficult and prolonged than those that served to find agreement on protecting the ozone layer →p.464.

1 WMO, *Report of the International Conference on the Assessment of the Role of Carbon Dioxide and of Other Greenhouse Gases in Climate Variations and Associated Impacts,* Geneva, 1986.

THE IPCC: A HIGH-LEVEL BALANCING ACT

The Intergovernmental Panel on Climate Change (IPCC) is the scientific authority on climate. The panel regularly summarizes the current state of knowledge as objectively as possible in extensive reports, collaborating with thousands of scientists from all over the world. As a body whose work lies between politics and science, it is expected to make "policy-relevant but not policy-prescriptive" statements. The IPCC faces enormous pressure and responsibility, and must comply with ever-stricter procedural rules.

The Intergovernmental Panel on Climate Change (IPCC) stands at the very center of climate negotiations, more or less a scientific conscience. The panel was first called together in 1988 by the World Meteorological Organization (WMO) and the United Nations Environment Programme (UNEP), and shortly after its first session it was endorsed by the UN General Assembly. Its mission is to provide an "internationally coordinated scientific assessment of the magnitude, timing and potential environmental and socio-economic impact of climate change and realistic response strategies." The IPCC was in fact not the first international consulting committee in the area of climate change, but it has developed into by far the most influential. Within the framework of the Climate Convention and the Kyoto Protocol it has the mandate—together with the subsidiary body for scientific and technological advice and the subsidiary body for implementation—to support parties to the convention in scientific and methodological matters.

Its chairmanship was given in 1988 to Swedish climate scientist Bert Bolin, who had gained a great deal of experience since the sixties serving on international committees that researched climate change. He had also coordinated climate research and had a good sense of what was feasible. He had been influential in preparations for earlier climate conferences, and most importantly had participated in the historic Villach conference in October 1985. This meeting had indirectly launched the initiative that later led to the creation of the IPCC, and indeed it was instrumental in kick-starting the call for a climate convention, which more and more voices rapidly joined →p. 452.

Mustafa Tolba, at that time the director of the UN Environment Programme and extremely active, first brought into being an Advisory Group on Greenhouse Gases (AGGG) in Villach in 1985 →p.219. In quick succession, this group organized further conferences that were supposed to bring the issue into the political arena. An international conference, "The Changing Atmosphere: Implications for Global Security," organized in Toronto in 1988 by a group of scientists and committed environmental policymakers, must also be indirectly counted among these. At this conference, following the example set in the 1987 Montreal Protocol on ozone-depleting substances, targets were proposed for reducing the emissions affecting climate that quite specifically called for a decrease of 20 percent by 2005 compared to 1988 levels.

However, the AGGG was neither well enough organized nor appropriately funded, nor was it sufficiently representative for its reports to make a decisive breakthrough. This was in sharp contrast to the scientific reports used to guide ozone policy in the Montreal Protocol, praised over and over again as a successful forerunner →p.452. UNEP and WMO, together with NASA, had, since 1981, regularly compiled baseline studies that summarized the state of science about the ozone problem, and that provided the basis for the Montreal Protocol. Later in an interview, Bolin, who belonged to the AGGG, said the reason for its limited effectiveness was that it had "no money and no muscles." But it was clear that it was even more important to have a strong and trustworthy scientific consulting group for guiding climate policy than was needed for tracking developments in the ozone layer. Up for debate, after all, were measures that would have much more comprehensive effects on the global economic system; added to that, the distinction between governmental and nongovernmental organizations at conferences had begun to blur in recent years. During and after the Villach conference, some scientists had begun to assume a more activist than scientific attitude.

Once it was established, the IPCC also benefited from the forerunner role of scientific ozone reports. Robert T. Watson of NASA, who had played a decisive role in mobilizing the fight to protect stratospheric ozone and who was a member of the executive committee for international ozone reports as of 1981, not only collaborated as principal author on the first IPCC Report of 1990; after Bolin's resignation in 1997, he also took over the chairmanship of the IPCC. Later, in 2002, at the instigation of the George W. Bush Administration, Watson was replaced by Rajendra Pachauri. Watson was considered a man of the Clinton presidency; Vice President Al Gore even called him his "hero of the planet." Watson had used forceful words to explain the findings in the third report to delegates at climate negotiations, and as an individual had spoken out frankly in favor of limiting greenhouse gas emissions.

The IPCC's mandate, consciously balanced between politics and science, was and is enormously difficult—the panel is expected to meet the highest standards in both areas. Climate science includes an almost unmanageable number of processes linked with each other in numerous ways; these are scrutinized in the most varied scientific disciplines, each with its own research traditions. At the same time, the findings of climate science carry implications of enormous consequence for the global economic system and society. The political and psychological pressure on IPCC experts is very high since the "course of the world" depends on their analyses. Their reports must be accurate and generally understandable. They represent a constant balancing act at the highest level. If they are formulated too carefully and become difficult to read, then they are not noticed and nothing happens. If they are too simplified, then they quickly create an alarmist effect. Both responses undermine trust in the institution.

The IPCC does not do any research itself, but at regular intervals summarizes the latest state of knowledge in assessments that are then adopted by consensus. These reports reflect the findings of three working groups. In the current fourth assessment (AR4), Working Group I looked at the physical science basis; Working Group II at impacts, adaptation, and vulnerabilities; and Working Group III at the mitigation of climate change. The three working group reports, published in 2007, together contain nearly 3,000 pages. In all, the reports comprise the work of 559 lead authors and 1,369 authors altogether, as well as 90,000 commentaries written by more than 2,500 reviewers. The IPCC also publishes, when needed, special reports on sectional aspects of the climate problem. Another important function is that the IPCC develops methods for estimating the quantity of greenhouse gas

The 2007 IPCC Report comprehensively documents current knowledge about climate change. It is based on thousands of research papers, about 100,000 data collections, continually improved climate model projections, and innumerable other model studies. For a period of almost six years, 1,369 authors from 130 countries and more than 2,500 experts contributed to this report.

emissions released. These methods are used in the preparation of national reports submitted in compliance with UN climate agreements. This is the only way to determine reliable and internationally recognized emissions values for individual countries.

The International Panel on Climate Change reached the high point of its career in December 2007 when Rajendra Pachauri, as IPCC chairman, accepted the Nobel Peace Prize in Oslo on its behalf. The panel shared the honor with Al Gore, former vice president of the United States. The Nobel Committee awarded them the prize for their "efforts to build up and disseminate greater knowledge about man-made climate change, and to lay the foundations for the measures that are needed to counteract such change."

Only two years later, however, the IPCC was caught up in its worst crisis, and disillusionment was high. Hackers stole e-mails from experts at the renowned Climate Research Unit at the University of East Anglia in Norwich in eastern England and posted them on the Internet. They wanted to prove that researchers involved to a considerable degree in the work of the IPCC, among them the head of the unit, were biased, had suppressed critical data, manipulated curves, and purposefully tried to exclude people who thought differently. Shortly afterward, accounts of a few errors in the fourth report cropped up. In particular, a statement that there was a high probability of glaciers in the Himalayas disappearing as early as 2035 proved to be scientifically unfounded. Late in January 2010, the IPCC had to officially confirm it had made that mistake, admitting that insufficiently proven estimates had been quoted from a popular science article and that criticism of this prognosis, drawing attention to its dubiousness, had not been taken seriously enough.

East Anglia University, UN Secretary-General Ban Ki Moon, and Rajendra Pachauri had the incidents investigated, and respectively the IPCC's working methods. None of the investigations called into question the IPCC Report's central statements, although it was noted that in the case of the e-mails there had been a lack of adequate openness. Indeed, one disputed chart really was misleading.[1]

In their report at the end of August 2010, the experts of the Inter-Academy Council mandated by Ban and Pachauri with examining the processes at the IPCC first underlined the importance of the panel.

However, they noted, among other things, that in the Working Group II summary for policymakers, statements were not sufficiently sup-port-ed by studies, put into context, or clearly formulated. Therefore they issued recommendations for fostering transparency and improv-ing communication, especially about uncertainties. The IPCC's man-date, they said, required that statements be "policy-relevant, but not policy-prescriptive." The experts also thought it appropriate to strengthen the leadership and the review process. They additionally recommend-ed that the term of office for the IPCC chair and the chairs of the working groups be limited to the amount of time needed for preparing one report.[2]

Shortly afterward, the IPCC decided to implement several of these recommendations right away. Intense debate about the other recommendations made by the InterAcademy Council took place at subsequent sessions and numerous measures to address the short-comings were approved. In the meantime, work on the fifth report has begun. It is expected to be finalized by the end of 2014 and will undoubtedly be inspected even more carefully than its predecessors were.

1 Muir Russell, *The Independent Climate Change E-Mails Review*, 2010.
2 Robbert H. Dijkgraaf et al., *Review of the Processes and Procedures of the IPCC*, Alkmaar, 2010.

RIO—KYOTO—MONTREAL

For more than twenty years, agreements for reducing global greenhouse gas emissions have been negotiated under the auspices of the United Nations. The UN Framework Convention on Climate Change laid down basic principles in a first step; in a second one, the Kyoto Protocol set specific targets. However, it became impossible to reconcile the interests of the three most important groups—Europe, the United States, and developing countries. The U.S. even refused to become a Kyoto state.

The Earth Summit of Rio de Janeiro—officially the UN Conference on Environment and Development (UNCED)—is seen by political observers as the greatest diplomatic event of the twentieth century. A total of 180 states sent diplomats to Brazil; 130 heads of state were in attendance, 10,000 journalists reported, and 17,000 people from all over the world took part from June 3 to 14, 1992. For the first time in the history of the United Nations, nongovernmental organizations were admitted and were even involved in preparatory work, although they did not have voting rights. The historic summit, occurring shortly after the fall of the Berlin Wall in 1989 and the breakup of the Soviet Union in 1991, was supposed to ring in a new era of global cooperation, above all between North and South, and consolidate environmental and development policy →p. 506.

Disillusionment has set in since then in many places, however, and the greening of the economic system has not made as much progress as was hoped for in Rio. Nonetheless, the Earth Summit marked a kind of awareness shift. Since then it has become generally accepted that responsibility for the earth must be shared and that historical development dictates that industrialized countries assume a greater load than developing countries. The Earth Summit in Rio triggered a great many processes at the most varied levels, in single towns, cities, countries, and nations, in politics, economics, and society. Many people in the South believe that the problems facing developing countries were given too little attention at the summit and afterward. The enforcement of the precautionary principle, the principle of common but differentiated responsibilities, and the linking of environmental and development policy has a long way to go. Numerous developing countries still have to fight hunger, extreme poverty, and high population growth. Access to clean water and basic health care is extremely precarious in many places, and more than sixty million children are still unable to attend basic primary school, almost half of them in Africa.

The United Nations Framework Convention on Climate Change was opened for signature in Rio de Janeiro in 1992. It went into effect barely three years later, in March 1994. In the meantime, 194 countries have acceded to the nearly thirty-page agreement; the latest country to join was Andorra in 2011. The ultimate objective of the convention is to achieve "stabilization of greenhouse gas concentrations in the atmosphere at a level that would prevent dangerous anthropogenic interference with the climate system." It also says, "Such a level should be achieved within a time frame sufficient to allow ecosystems to adapt naturally to climate change, to ensure that food production is not threatened and to enable economic development to proceed in a sustainable manner."

The principles laid out in the first parts of the convention are just as important. The parties to the convention declare their intention to protect the climate system for the benefit of present and future generations "on the basis of equity and in accordance with their common but differentiated responsibilities and respective capabilities." Additionally, "where there are threats of serious or irreversible damage, lack of full scientific certainty should not be used as a reason for postponing such measures." Full consideration should be given to the special needs of developing countries, above all the very vulnerable ones. In any case, the convention also says, "policies and measures to deal with climate change should be cost-effective so as to ensure global benefits at the lowest possible cost."

The industrialized states have special obligations since they generated the greatest share of earlier emissions as well as most of those measured in 1990, but also because "per capita emissions in developing countries are still relatively low" and their emissions should be allowed to go up to cover their development needs. The industrialized countries are listed by name in Annex I of the convention. They are required to define national strategies and install measures to indicate

“The impacts of climate change are not evenly distributed—the poorest countries and people will suffer earliest and most. And if and when the damages appear it will be too late to reverse the process. Thus we are forced to look a long way ahead.”

 The Stern Report, compiled by Nicholas Stern for the UK government in 2006 that they will be the first to start reducing emissions and thereby show the way forward. The overall goal of the convention is to reduce emissions to 1990 levels. It is not only about the greenhouse gases generated by human activity, but also about protecting and strengthening greenhouse gas sinks and reservoirs like forests and soils →p. 518. Finally, industrialized countries are supposed to rapidly and regularly publish information on their efforts. On the basis of these reports, the parties to the convention are to decide whether further steps are needed, including additional obligations for industrialized states.

The industrialized countries of the West are listed separately in Annex II. They have promised to make money available to compensate developing countries for the costs incurred from observing the convention and meeting its obligations. Additionally, poor countries that will be negatively affected by climate change are to be aided in making adaptations. Measures are also foreseen to enable the easy transfer of climate-friendly technologies to developing countries. The industrialized countries in the former Eastern Bloc are, however, exempted from these financial obligations, and they are also allowed flexibility in meeting their Annex I targets.

The convention explicitly states that developing countries need to fulfill the obligations assigned to them within the framework of the convention only to the extent that industrialized countries in turn meet their obligations in financing and technology transfer. For developing countries, it is the general obligations of the convention that are in the foreground. These include compiling national inventories of greenhouse gas emissions, designing and implementing programs to mitigate climate change by reducing emissions and protecting sinks, and educating and informing their populations about the climate problem.

A lengthy political tug of war preceded the acceptance of this convention text even before actual negotiations began in 1991. Various key points of the framework agreement were defined beforehand, for instance at the second World Climate Conference in 1990 in Geneva →p. 452. Political alliances and contentious points remained the same, however, and later also characterized the work on a climate protocol tackled in 1995 at the first Conference of Parties in Berlin (COP1) and its further fate.

Let us look at the position of the United States, for instance. Starting in the fall of 1990 in Geneva, and later in the Intergovernmental Negotiating Committee, the U.S. was always very interested in taking sinks and reservoirs of greenhouse gases into account. The capacity of forests to absorb carbon dioxide from the atmosphere was to be included in calculations, as was logging, which releases large amounts of greenhouse gas. Finally in 1997, when the first climate protocol, now well known as the Kyoto Protocol, was being formulated at the third Conference of Parties (COP3) in Kyoto, American negotiators were able to assert this interest in a fierce battle against both European and developing countries. Likewise, they wanted to discuss not just carbon dioxide emissions and therefore insisted on a more comprehensive solution, stressing that other, extremely potent greenhouse gases should also be regulated. These gases were not of any great consequence at the time, but were projected to see rapid growth. Finally, the United States also wanted to be able to meet its obligations beyond its own borders so that it could reduce emissions wherever this would cost the least, intending to keep its overall economic burden as low as possible. This idea led in 1997 to the Kyoto mechanisms, also called flexible mechanisms or flexibility mechanisms →p.476. European countries, on the other hand, led by the European Union and supported by many environmental organizations that had worked hard to raise public awareness, saw success lying primarily in reducing domestic emissions. Policies and measures were very high on their agendas. These were meant to promote technologies to increase energy efficiency and to meet energy needs from renewable energy sources such as the sun, water, wind, biomass, and so forth. Policy ideas focused on obliging individual countries to decrease the volume of their greenhouse gas emissions, a concept discussed under the heading of QELRO (Quantified Emission Limitation and Reduction Objectives). As early as the World Climate Conference in 1990 in Geneva, European countries would have liked to set specific targets for stabilizing emissions at 1990 levels, and they wanted to set them explicitly as a political goal for the UNFCCC too.

This kind of reduction was easier for the EU to achieve however than it was for the United States. Because the European Union insisted on being handled as a single unit, it was able to distribute responsibilities among its member states at its own discretion, even without flexibility mechanisms. Added to that, the reunification of Germany meant that there was an unusually large potential for emission reduction that could be realized at low cost by shutting down obsolete plants in former East Germany. The EU could also benefit from the fact that Britain had switched to generating electricity with natural gas instead of coal in the 1990s, which lowered CO_2 emissions. European negotiators, however, criticized the United States' interests, seeing them as loopholes that allowed the American economy to carry less of a burden. They were equally critical of the option for meeting reduction targets outside of domestic borders, and for including forests in calculations. Forests in the United States held the promise of being reflected as large sinks in the emissions budget.

From the very beginning, perceptible areas of tension also arose between the United States and emerging countries, principally China—and were still present at the 2010 climate conference in Cancún twenty years later. The coalition of developing countries known as the Group of 77 and China categorically resisted any wording that would have saddled them with any additional obligations, referring back to the UNFCCC and "differentiated" responsibilities. American negotiators, in contrast, pointed early on to rapidly increasing emissions in emerging countries and called for comprehensive regulations. It was this problem most of all that led to the fact that no American president has ever submitted the 1997 Kyoto Protocol, with its quantitative reduction guidelines, to the U.S. Senate for ratification, even though the United States had ratified the rather generally formulated UN Framework Convention on Climate Change negotiated five years before the Kyoto conference. This was also the case for the Clinton Administration, which flew vice president Al Gore to Japan in 1997 to calm down the furious mood during negotiations in Kyoto. Gore, who had campaigned in 1992 with his book *Earth in the Balance,* was considered a great beacon of hope among environmental activists.

But environmental activists and Europeans at Kyoto were obviously slow to realize that Gore and his president had little leeway. In the run-up to the conference, and alarmed by the negotiation mandate for the protocol adopted in Berlin in 1995, the United States

Senate had already made it clear in the summer of 1997 in the Byrd-Hagel Resolution that a protocol in the framework of the UNFCCC had to meet certain conditions. The first condition was that the protocol had to set targets for important emerging countries to limit or reduce greenhouse gas emissions within the same commitment period; and the second condition was that it must not inflict any serious damage on the American economy. The resolution was meant to prevent American acceptance of the Kyoto Protocol if it allowed countries like China, Mexico, India, Brazil, and South Korea to be counted among the group of developing countries that did not have to comply with meeting specific reduction goals. The resolution expressed the expectation that emissions in these emerging countries would exceed those of the United States and other OECD countries as early as 2015. After detailed debate, the U.S. Senate passed this resolution at the end of July 1997 without a single opposing vote. Already at the Kyoto conference, these conditions made it essentially clear that the negotiated compromise would not have a chance of being formally adopted by the United States.

President Clinton and Vice President Gore had the same experience that George Bush Senior had had eight years earlier. Bush, at that time the Republican presidential candidate and vice president under Ronald Reagan, explained in an environmentally groundbreaking speech in August 1988 that anyone who thought people were powerless to do anything about the greenhouse effect was forgetting about the "White House effect." He said that as president he intended to take action. He called for a worldwide environmental conference and promised to attend it personally. Although he quickly kept his promise to renew the Clean Air Act and saw a draft become legislation that had been written partly by committed environmental politicians, and although he traveled to the Rio Earth Summit himself, he ran into unexpectedly high hurdles when it came to the climate. Members of his administration later thought that he had underestimated the economic challenges linked to an active policy for mitigating climate change and consequently underestimated the resistance posed by special interests.

The resistance that had already shown itself early in the nineties within the administration was probably rooted in several factors.

Although American scientists were among the pioneers in research on the greenhouse problem and climate science was supported early on with a great deal of money, doubt about climate change had always been expressed in the United States. Besides that, the U.S. has very high per capita consumption levels, and the average American in 1990 was responsible for emissions of almost 20 tons of carbon dioxide, about twice as much as an inhabitant of Britain or Japan, nearly nine times more than an inhabitant of China, and twenty-two times more than an inhabitant of India. The global average of CO_2 emissions per person at that time was about 3.9 tons. Per capita emissions of CO_2 in Benin, Niger, Mozambique, Burkina Faso, and Bangladesh were only about 0.1 tons.

It is not just that the United States is a rich country. The infrastructure of American society in most regions relies heavily on cheap fuels, today mostly on oil. Towns, cities, and metropolitan areas are spread out over enormous areas, and public transportation is almost completely nonexistent in many places. Cheaply built single-family homes with poor insulation and high energy needs for heating and cooling are predominant. A lifestyle with huge shopping malls, large refrigerators, and all kinds of automated household appliances contributes to this thirst for energy in the land of opportunity. At the same time, most Americans are so preoccupied with their own economic advancement and their personal surroundings that they give little attention to international issues.

The political system also differs from European models. The American president admittedly has considerable power, but he must also take into account the checks and balances anchored in the U.S. Constitution. Elected congressional representatives in the House of Representatives and the Senate, and lobbies, have great influence on policymaking. And the pioneer spirit is still invoked in this country. The courage to take risks, be adventurous, and surpass limits carries more prestige than being concerned about safety and security. The European way of thinking about policies and measures does not correspond to an American "philosophy" on the environment, as American representatives had tried to explain much earlier at the World Climate Conference in Geneva in 1990. Later, under Clinton and Gore, American negotiators at climate conferences again referred to these differences in thinking, propagating voluntary action and counting on the development of new technologies and market mechanisms.

The third strong negotiating partner at the table is an organization of developing countries, the Group of 77 and China, as they call themselves within the framework of UN negotiations. For them, the negative consequences of climate change (caused by industrialized states) are in the foreground. The poorer a population, the more defenseless it is, left at the mercy of droughts, severe weather, hurricanes, and floods. At the same time, these countries are also concerned about their own development. Both the need to protect themselves from the negative consequences of global warming and to overcome poverty requires money and technology. Many of the 2015 Millennium Goals, aiming to improve the situation in poor countries, are still very far from seeing achievement.

The more than 130 developing countries, among them the emerging economies, carry great weight as a group. They do have a broad range of interests however. Here China, the heavyweight, must be mentioned first. In recent years it has become the biggest emitter of greenhouse gases in the world, and its rapidly growing economy has turned it into a significant competitor of the United States. The government in Beijing reacts just as sensitively to attacks on its own sovereignty as does the government in Washington, D.C.

There are other rapidly emerging states in the Group of 77 and China, among them Brazil, India, Argentina, Chile, and South Africa. Singapore, whose per capita income exceeds that of various countries in the EU, and rich oil countries like Qatar, the United Arab Emirates, Kuwait, and Saudi Arabia, are in this group as well. It also includes countries in the very poor regions of the Sahel and sub-Saharan Africa, as well as Bangladesh, Bhutan, and Vietnam, and small island nations, such as the Maldives, Kiribati, and parts of the Bahamas, which are especially threatened by a rise in sea levels and the consequences this brings. All these countries in the Group of 77 and China clearly have very different priorities. Nonetheless, in climate negotiations, they take their positions in a united front against industrialized states—or at least that was the case until recently →p.500. This allowed even oil-producing countries, especially dependent on income from fossil fuels, to be listed in a group whose special needs and interests are

"Poverty and climate change are the two great challenges of the twenty-first century. Our responses to them will define our generation, and because they are linked to each other, if we fail on one, we will fail on the other."

 Nicholas Stern, economist, 2009

particularly important. All these countries share the advantage that the UNFCCC and the Kyoto Protocol do not pose an extra burden to them. They are highly interested in not fundamentally changing this situation, and to make their argument they refer to their need for development and to the historical responsibility of industrialized states.

However, without a significant contribution from emerging countries, no movement was to be expected in the United States. In March 2001, a few weeks after his inauguration as president, George Bush Junior therefore unceremoniously put an end to the debate over further development of the Kyoto Protocol, pending at that time and so unpopular in the Senate. He announced that no targets for greenhouse gases would be enacted and characterized the Kyoto Protocol as "fundamentally flawed." Like his father before him, he had made promises before entering office that he did not keep.

Bush's decision not to join the Kyoto Protocol in 2001 provoked outrage not only among Europeans, environmental activists, and developing countries, it also led to a situation in which it was doubtful for a long time whether enough industrialized states would accede to it so that it could enter into force →p.476. Canada, Australia, and Japan, for example, were especially close to the United States, and had joined together with the U.S. and others in the so-called Umbrella Group. Until May 2002, it was unclear whether or not Japan and its prime minister Koizumi would come down on the side of the United States. Even Russia had to be asked for a long time to join, but after several concessions were made, it did finally ratify the Kyoto Protocol on November 18, 2004. This enabled the protocol to go into force ninety days later, just in time for member states to begin preparing for the first commitment period, which was to start barely three years later. For the first time, specific, quantitative targets for reducing greenhouse gas emissions were defined and categorized as internationally binding. However, the final step of recognizing compliance rules, which must be ratified by individual states, and which specify that failure to comply with obligations meets with disciplinary measures, had still not been taken by 2011.

Nine months after the protocol went into effect, Kyoto Protocol signatories met in Montreal in 2005 for the first official conference of parties. The agreement had required industrialized states to verify emissions limits and reductions for the commitment period from 2008 to 2012, and to begin preparations in 2005 for a second commitment period. This raised the awkward question in Montreal as to whether European states, together with Japan, Russia, and Canada, were willing again after 2012 to reduce greenhouse gas emissions in an internationally binding and verifiable way—without the United States and without major emerging countries, more or less on their own. George Bush and the United States, now at war in Afghanistan and in Iraq after the events of 9/11, had long since become the bogeyman in many places. Nonetheless, Japan, for instance, showed little desire to go on without the U.S.—in other words, without a truly global agreement that the United States would also join. It voted for talks outside the Kyoto Protocol conference but within the overarching framework of the UNFCCC. This would enable the inclusion of countries that had not signed the protocol. Japan let it be known that it saw the Kyoto Protocol as an initial step only.

After difficult negotiations, which the U.S. temporarily left in protest, parties in Montreal in 2005 finally agreed, under the umbrella of the climate convention, to carry on a two-year dialogue about long-term cooperation. However, American negotiators insisted that it should be explicitly "non-binding" and "not open up any negotiations that would lead to new obligations."

It still looked as if there was enough time to reach a settlement for the period after 2012.

THE KYOTO PROTOCOL

The Kyoto Climate Protocol has shaped international climate policy in recent years like no other international agreement. Industrialized countries, with the exception of the United States, committed themselves to reducing their greenhouse gas emissions from 2008 to 2012, the first commitment period, by at least 5 percent. It was only at the very end of 2011 when it became clear that there would indeed be a second commitment period.

There is hardly an international agreement in the last few years that has received as much public attention as the Kyoto Protocol, which has been at the center of global climate policy since the second half of the 1990s. A good twenty pages long, its text was unanimously adopted at the third Conference of Parties (COP3) of the UN Framework Convention on Climate Change—the official name of the conference in Kyoto in December 1997—by the member states of the climate convention. The wording of the agreement was subject beforehand to fierce fighting and long negotiations, and it would be another four years before it was really clear what the rules were →p. 464. Only at COP7 in Marrakesh in 2001, and after the United States had definitively turned its back on the protocol, was it possible to reach agreement on essential principles for interpreting and applying the Kyoto Protocol. These were laid out in rulings that covered more than three hundred pages, the so-called Marrakesh Accords. Further clarification followed. Another three years and many a political tug of war later, the agreement finally went into force in February 2005. Admittedly, the reduction in emissions aimed for is generally low in view of the magnitude of global emissions and greenhouse gas fluxes, but the idea was that it would at least force an initial revision in thinking. Altogether, 192 countries have acceded to the agreement in legally valid form.

Comprising twenty-eight articles, the Kyoto Protocol is shaped by asymmetry—there are two classes of parties to the agreement, meaning that 37 industrialized countries face more than 150 states in the developing world. The industrialized countries are listed in Annex B of the protocol, and have accepted binding quantitative obligations for reducing their emissions of six greenhouse gases. Emissions of carbon dioxide, methane, nitrous oxide (laughing gas), hydrofluorocarbons, perfluorocarbons, and sulfur hexafluoride are all supposed to be jointly reduced by at least 5 percent in comparison with 1990 levels (in some cases with 1995 levels) during the commitment period from 2008 to 2012. In contrast, the protocol explicitly states that developing countries, in recognition of "common but differentiated responsibilities" have no further obligations other than those already mentioned in the UNFCCC of 1992. In the foreground of this document are programs for mitigating climate change and adapting to its consequences, as well as guidelines for reporting.

The asymmetry of obligations was compensated to some extent by an asymmetry in the conditions to be met for the protocol to enter into force. Not only fifty-five parties to the agreement were needed for that, but also a sufficient number of industrialized states (the countries listed in Annex I of the UNFCCC) had to definitively sign the protocol so that their carbon dioxide emissions encompassed at least 55 percent of the total emissions in 1990 as listed in Annex I. The United States' refusal to accede therefore posed a major problem for the ratification process, since the U.S. alone at that time accounted for 36.1 percent of Annex I emissions. If Russia, with a 17.4 percent share of emissions, had likewise turned its back on the protocol, the emissions of all remaining countries together would not have been enough to meet the qualification posed by the 55-percent clause. The protocol would have been rendered obsolete.

Not all industrialized states, however, have to lower their emissions by the same percentage (in comparison with 1990 levels) during the commitment period. Iceland, Australia, and Norway were even allowed a slight increase. Japan, on the other hand, accepted a reduction target of 6 percent, the United States signed in Kyoto for a target of 7 percent, and most European countries agreed to 8 percent →p.496. These differing figures were an outcome of the search for consensus in Kyoto. Initially there had been agreement on a common target of 5 percent; later, the chair of the conference had each state declare its own goal and then added all reductions together. The total was approximately 5.2 percent and led to the formula of "at least 5 percent" that was set in the protocol. The industrialized states are supposed to meet their targets mostly through national programs and measures. They are also obliged to report annually on the success of their efforts.

In the view of environmental activists and many Europeans, however, the decision to set actual quantitative reduction targets (QELROs, formally "quantified emission limitation and reduction objectives") had its price. The United States, which had insisted early, albeit unsuccessfully, on target obligations for rapidly emerging countries, agreed only conditionally to QELROs →p.468. One of these conditions was to have different options for achieving reduction targets, in particular market instruments. This resulted in the acceptance of market-oriented Kyoto Mechanisms, also called flexibility mechanisms or flexible mechanisms. Admittedly, the policies and measures called for by Europeans are in the protocol too, as vehicles to be implemented and further elaborated to achieve reduction targets. The central element of the protocol, in addition to national programs, is however the possibility of meeting reduction obligations outside domestic borders, mainly through emissions trading, a taboo subject during many of the negotiations.

Emissions trading enables industrialized countries to sell each other their emissions rights, termed assigned amount units (AAU). One AAU stands for the emission of one ton of CO_2 or the CO_2 equivalent of another greenhouse gas. The European Union has also developed its own market for such CO_2 emissions, the European Emissions Trading System (ETS) →p.401. Industrialized countries were also granted the opportunity to earn emission reduction units (ERU) through individual projects in other Annex B countries of the Kyoto Protocol. States in the former Eastern Bloc, transition countries in which the market economy was just beginning to take hold at the time of protocol negotiations, are especially suitable for such endeavors, referred to in specialist jargon as joint implementation projects.

Finally, under the Clean Development Mechanism (CDM) created for this purpose, wealthy countries can obtain Certified Emission Reductions (CERs) by means of climate-friendly projects in developing countries. Two percent of these CERs must be paid into a fund, the Adaptation Fund, set up especially to help vulnerable developing countries adapt to climate change →p.506. However, only those Annex

B countries that meet their obligations within the scope of the protocol are allowed to participate in trading with such Kyoto units.

But it was not only the introduction of market mechanisms that met with opposition among Europeans and climate activists in Kyoto negotiations. The same is true for the fact that, at the urging of American negotiators, forests as sinks of greenhouse gases as well as other such sources and sinks can be included in calculations. The acronym LULUCF (land use, land-use change, and forestry) represents this extremely complex subject, which after Kyoto still had to be analyzed in detail and put into specific terms during many years of negotiation. The performance of sinks that extract greenhouse gases from the atmosphere is likewise included in the balance of allotted emissions rights, in the form of removal units (RMUs) →p. 518.

The protocol distinguishes between two categories of sources and sinks. For the first it is a matter of "direct human-induced land-use change and forestry activities" that should be taken into account in a way "limited to afforestation, reforestation, and deforestation since 1990." The second category includes additional measures in the area of "agricultural soils and land-use change and forestry." They are to be taken into account by the parties starting in the second commitment period, although countries can have them taken into account already during the first commitment period if they choose to do so and if they are declared as such in time. In contrast to other Kyoto units, RMUs cannot be transferred to a later commitment period.

In 2012, it could still not be definitively said whether the Kyoto Protocol target to decrease emissions by at least 5 percent compared to 1990 levels (equivalent to 2.6 billion tons of CO_2) would be achieved during the five years of the first commitment period. According to the 2010 figures from climate convention experts, emissions in 2008 from all industrialized states in the convention—that is, all countries in the Kyoto Protocol that had reduction targets—were around 6.1 percent below 1990 values, without taking sinks (LULUCF) into account; if LULUCF are included, then the figure is around 10.4 percent. This pleasing result was primarily due to former Eastern Bloc countries releasing 36.8 percent less in emissions than they did in 1990 (if LULUCF were taken into account, the figure would be 48.5 percent). Emissions in Turkey, Iceland, Spain, Portugal, Australia,

Canada, Greece, Ireland, and New Zealand increased especially sharply, namely between 96 and 22 percent (without LULUCF). The United States also saw 13.3 percent growth in emissions, representing a very large quantitative increase of greenhouse gases. On the other hand, Britain and Germany featured the largest emissions reductions among Western industrialized states, with a decrease of approximately 20 percent.

No reduction targets were set in Kyoto in 1997 for later commitment periods, and it was not possible to reach a final agreement until December 2011, only a year before the end of the first period, in spite of intensive negotiations →p.500. Therefore, after the debacle at the 2009 Copenhagen Climate Change Conference, attempts were already being made to prevent a "commitment gap" occurring at the beginning of 2013, since such a gap would have negative repercussions on the entire agreement and in particular on the market with CO_2 certificates. The 2011 Durban Climate Change Conference was once again unable to work out all the details for regulating a second commitment period, but at this meeting the Kyoto parties decided that there will be a next commitment period starting January 1, 2013, covering the next five (or, alternatively, eight) years. Thus there will be no gap, and continuity is guaranteed even though not all the former Kyoto countries will participate. Russia, Canada, and Japan had already made it clear long before that they did not intend to participate in a second commitment period.

This does not mean, however, that these countries are refusing in principle to accept limitations on emissions, although they differ in ambition. But they do want to set their targets within the framework of a new agreement, a global climate change regime that has been under negotiation since the 2007 Bali conference, in parallel with the settlement of a second commitment period in the Kyoto Protocol. The intention of a new global climate regime is to see that binding reduction targets are also set for large emerging countries and the United States. Declarations of intent for relevant emissions reductions are noted in the Copenhagen Accord, the final document resulting from the conference in Copenhagen. In Durban, the parties of the climate convention decided that such an agreement shall be finalized no later than at the climate conference at the end of 2015 and enter into force in the year 2020. Most remarkably, the old division between developed and developing countries is not mentioned.

With respect to the progress made on the Kyoto Protocol in Durban, however, the Ad Hoc Working Group on Further Commitments for Annex I parties under the Kyoto Protocol has been asked to finalize open questions as soon as possible and, above all, convert existing pledges for emission reductions into QELROs for adoption at the end of 2012, at the next climate conference in Qatar.

June 3, 1992. Optimism after the Cold War—development and environmental issues are at the heart of international politics for the first time at the Earth Summit in Rio de Janeiro. *Pendergast/UN Photo*

The international community recognizes climate change as a problem; Brazil's President Fernando Collor de Mello signs the UN Climate Convention at the Earth Summit. *Eduardo DiBaia/AP/Keystone*

Beginning of negotiations on specific reductions in greenhouse gas emissions — the first UNFCCC Conference of Parties (COP1) opens in Berlin in 1995. *Jan Bauer/AP/Keystone*

Angela Merkel, as Germany's environment minister and first president of a COP, accepts the gavel in 1995 to officially chair the opening session in Berlin. *Jockel Finck/AP/Keystone*

Playing ball with the Earth — a WWF masquerade action at the 1997 climate conference in Kyoto urged leaders to take decisive action. *Eriko Sugita/Reuters*

A beacon of hope — U.S. vice president Al Gore, here with Japan's prime minister Hashimoto and foreign minister Ouchi, made a personal appearance at the 1997 climate conference in Kyoto. *Katsumi Kasahara/AP/Keystone*

Drafting the Kyoto Protocol involved night-long negotiations to the point of exhaustion. Some talked behind closed doors; others had to wait. *Katsumi Kasahara/AP/Keystone*

A first breakthrough—chief negotiator Raul Estrada Oyuela, right, announced on December 11, 1997 that delegates had accepted the Kyoto Protocol. *Mainichi Shimbun/AP/Keystone*

Disagreement among states was still substantial in November 2000 and the Kyoto Protocol was far from ratification. Demonstrators in The Hague called for action. *Fred Ernst/Reuters*

Waiting for a breakthrough at the November 2000 climate conference in The Hague. The conference broke up after two weeks with no outcome and was continued in Bonn in 2001. *Fred Ernst/Reuters*

Anger at U.S. president George Bush culminated at the July 2001 climate conference in Bonn. Bush had clearly said no to the Kyoto Protocol a few months earlier. *Hermann J. Knippertz/AP/Keystone*

Environmental organizations called for prompt ratification of the Kyoto Protocol during the Bonn climate conference in 2001. *Hermann J. Knippertz/AP/Keystone*

The Kyoto Protocol was finally ready for signing at the October 2001 conference in Marrakesh (COP7). The United States was no longer part of it. *Michel Euler/AP/Keystone*

Former U.S. president Bill Clinton in Montreal for COP11, presenting ideas for local measures as the American way to reduce greenhouse gas emissions. *Ryan Remiorz/Keystone*

The Kyoto Protocol went into force on February 16, 2005—without the United States, which regarded limiting greenhouse gas emissions as detrimental to its economy. *Jason Reed/Reuters*

UN Secretary-General Ban Ki Moon urges the plenary of the Bali climate conference (COP13) to find agreement. The Bali Action Plan lays the foundation for future climate negotiations. *Ed Wray/AP/Keystone*

Rajendra Pachauri, on behalf of the Intergovernmental Panel on Climate Change (IPCC), and Al Gore share the Nobel Peace Prize on December 10, 2007 in Oslo. *Bjorn Sigurdsen/Keystone*

The 2008 collapse of investment bank Lehman Brothers triggered a global crisis, raising a hurdle for the transfer of funds to developing countries for climate protection measures. *Mary Altaffer/Keystone*

U.S. president Obama and EU leaders at the 2009 climate conference in Copenhagen (COP15), searching for common ground. *Ho New/Reuters*

Hopes were high before the 2009 conference in Copenhagen (COP15) that a new regime would be found that committed the United States and emerging countries to reducing emissions. *Bob Strong/Reuters*

Bolivian women demand rights for nature in climate policy at the World People's Conference on Climate Change and the Rights of Mother Earth. Cochabamba, Bolivia, April 2010. *David Mercado/Reuters*

Mexico's president Felipe Calderón speaks at the climate conference in Cancún (COP 16) on December 11, 2010 to delegates who have just settled on the Cancún Agreements. *Bao Feifei/Keystone*

The environmental organization 350.org calls on delegates at the Cancún conference (COP16) in December 2010 to limit global warming to 1.5 degrees Celsius (2.7°F). *Jorge Silva/Reuters*

1.5 ⁰C
To Stay
ALIVE
1.5 ⁰C
To Stay
ALIVE
350ppm
=
SURVIVAL
ippines
Dominican
Republic
Mexico
Kiribati
COP16
CMP6
México2010

HOT AIR

The language used at United Nations climate conferences is full of odd abbreviations and acronyms. There is talk of LULUCF, QELRO, CDM, JI, AAU, CER, REDD+, MRV, and much more. This "insider slang" reflects the complexity of negotiations in which new issues are taken up over and over again and defined and redefined, and in which new ways to solve problems must be found. The same holds true for "hot air," an unofficial term coined by environmental organizations.

Among the strange blooms that flowered in the political greenhouse of the 1997 Kyoto Climate Change Conference →p.464 was the term "hot air." Hot air is what environmental organizations call the excess of greenhouse gas emissions rights assigned to former Eastern Bloc countries, rights created by setting easily achievable targets. The emission limits of these so-called economies in transition are unrealistically high. Even without making the slightest effort toward adopting stringent climate policies, they will still have a large quantity of unclaimed emissions rights left over at the end of the 2008–12 first commitment period. Emissions trading regulations anchored in the protocol stipulate that excess emissions rights, also called surplus assigned amount units (AAU), can be sold to other countries →p.401. As green critics therefore often note, the Kyoto Protocol has made it possible for former Eastern Bloc countries to earn money with hot air.

According to the Kyoto Protocol, Russia is allowed to emit exactly as much greenhouse gas during each year of the first commitment period as the Russian Soviet Socialist Republic emitted in 1990. Moscow set this target for itself in Kyoto, just as other countries did, with the difference however that most other states set themselves targets to lower emissions by a few percentage points. Emissions in Japan and Canada, for instance, are supposed to be 6 percent lower, and those of the European Union and Switzerland 8 percent lower than in 1990. But according to the Kyoto Protocol, Russia is still entitled to emissions of 3.32 billion tons of CO_2 equivalents annually throughout the entire commitment period from 2008 to 2012.

Not only is Moscow's target much less ambitious compared to targets in many industrialized states—the breakup of the Soviet Union, with all its economic problems, has also led to greenhouse gas emissions in Russia falling off sharply after 1991 without any government involvement. Russia's emissions—after the EU, it is the second largest emitter among the industrialized states that joined the Kyoto Protocol—were nearly 40 percent lower in 1997, when the protocol was negotiated, than in 1990. It was already clear in Kyoto that modernization of the dilapidated industrial sector would lead to considerable improvements in energy efficiency, making it possible to increase industrial performance without burning additional fossil fuels. Moscow could never exhaust its contingent of emissions units in the first commitment period by 2012 and would be able to benefit from windfall profits. EU experts calculated that during the five years of the first commitment period, Russia would have more than 5.5 billion tons of CO_2 equivalents, or hot air, at its disposal. This is almost six times the volume of greenhouse gas emissions that Germany released in all of 2007 (just under 960 million tons).

Russia in fact is the most important, but it is not the only state that created hot air for itself by setting a less than ambitious emissions target, allowing it to theoretically sell emissions rights to other industrialized countries. The same is true for Ukraine, which is also allowed to emit exactly as much each year as in 1990, although by 1997 its emissions had decreased by more than half, and ten years later were still under 50 percent of the 1990 level. Even the European Union, through its accession of states in the former Eastern Bloc, has hot air at its disposal. During the commitment period from 2008 to 2012, the EU was allowed to accumulate probably 2.2 to 2.5 billion tons of CO_2 equivalents, an amount similar to that of Ukraine. If the hot air for all industrialized countries that signed the Kyoto Protocol is added up, it will reach a volume of 10 to 13 billion tons in this period. This is many times more than the 2.6 billion tons by which all Kyoto industrialized states together were supposed to decrease their emissions during the five-year commitment period. Or in other words, if circulation of this hot air could be successfully prevented, the reduction effect achieved in this way would exceed by many times the reduction volume that the other Kyoto signatories are expected to achieve.

Within the framework of European emissions trade, however, there are already certain restrictions on EU countries using such hot air. And outside of the European Union, it is also no longer so easy for transition countries to sell hot air in the way originally feared by environmentalists in Kyoto. Buyers of such surplus assigned amount units have no interest in seeing the Kyoto Protocol undermined in this way. Most of them attach conditions to their purchases and require sellers to invest their profits in projects, technologies, or programs that will lead to genuine reductions in greenhouse gas emissions in the medium term. Such contracts are implemented under Green Investment Schemes (GIS) that call for the establishment of appropriate institutions in transition countries; they also stipulate that compliance is monitored by independent bodies.

Nonetheless, the subject of hot air is still present, even years after the Kyoto conference as well as after the conferences in Copenhagen, Cancún, and Durban. It concerns the future. The decisive factor will be how rules are set for transferring these emissions rights from the first commitment period of the Kyoto Protocol to the second. However, the fact that Russia made an official statement in Cancún that it did not intend to set targets for a second commitment period has significantly defused the problem. Without Moscow, almost half of the hot air in the Kyoto system dissipates →p. 500. For now, what happens to the rest of the hot air at the end of 2012 is still an open question. The text of the agreement adopted in December 2010 at the climate conference in Cancún, as the basis for further shaping of the Kyoto Protocol, still held various options. The Climate Change Conference in Durban a year later thus directed the relevant working group to quickly clarify the role hot air can play for the reduction goals of the second commitment period in order that the results be taken into account when formulating the new QELROs, which are to be established in Qatar at the end of 2012.

The 2007 IPCC Report found it *very likely* (greater than 90 percent chance) that most of the warming observed in the past fifty years is due to the increase of greenhouse gas emissions from human activities. The prior report in 2001 had deemed this *likely* to be the case (greater than 66 percent chance).

THE FUTURE BEGINS IN BALI

The Kyoto Protocol of 1997, the first climate agreement with quantitative targets for reducing greenhouse gas emissions, has not met all the hopes placed on it. The United States, still the biggest emitter among industrialized states, is not part of it. Emissions in developing countries have more than doubled since 1990, and China is now emitting more greenhouse gases than any other country. Work on a new, truly global agreement has been proceeding for several years—but old conflicts of interests have impeded progress. Yet there is a pledge that an inclusive solution, "an agreed outcome with legal force" applicable to all parties, will come into effect in 2020.

In the afternoon of December 15, 2007—after one of those notoriously long nights at the end of a climate conference, with hours of waiting while individual countries came to agreements within their own small groups—the breakthrough came during a highly emotional plenary session. The applause of delegates at the Nusa Dua conference center in Bali was enthusiastically loud when the worn-out conference leadership finally announced with relief the acceptance of the Bali Action Plan to the 13th Conference of Parties (COP13) of the UNFCCC. Shortly beforehand, UN Secretary-General Ban Ki Moon had once more urgently admonished delegations to make use of this historic opportunity for an ambitious agreement. Only a few journalists appreciated the result however; they had set the bar too high, and everyone was too preoccupied with working out guidelines for the Kyoto Protocol beyond 2012. But COP13 had really rung in a new round of negotiations. Within two years it was expected to work out a completely new climate regime with emissions targets for the industrialized signatories of the Kyoto Protocol as well as comparable targets for the United States and substantial obligations for emerging countries. The Bali Action Plan laid down the central objectives for this new global climate regime; the work was to be finalized in Copenhagen in 2009 and submitted to the plenary session for acceptance.

Two years earlier, shortly after the Kyoto Protocol had gone into force in 2005, negotiators had already begun to give thought to the time after 2012 →p.464. Indeed, the protocol envisaged this. The first commitment period of the protocol, with specific reduction targets for participating industrialized states, would run out in 2012. Therefore, after the Kyoto agreement went into force there was little time left for celebrating, and it was apparent that upcoming negotiations for a second commitment period would not be simple. But if an emissions regime like the Kyoto Protocol did not include the biggest emitter among industrial countries, the United States as world leader, it could only be a partial success. And the Group of 77 and China, the union of developing countries, referring to their great need to catch up economically, were still not willing to be bound to specific quantitative targets. Accordingly, there was no solution in sight to resolve the American demand that powerful emerging countries, still counted among developing countries, also take on reduction obligations. Therefore no one was really counting any more on the only remaining superpower to finally accede to the agreement. The United States had no longer participated in negotiations over the Marrakesh Accords as of 2001, nor in later work on details of the Kyoto Protocol—a fact that may actually have facilitated their completion →p.476.

Japan and Switzerland, for instance, no longer wanted debate on further obligations to be limited to Kyoto Protocol signatory states. This quickly became clear in 2005 in Montreal at the first official meeting of parties to the Kyoto agreement. Bush Administration negotiators, on the other hand, pilloried for a long time as the black sheep of climate policy, did not want to enter binding talks or any dialogue that could lead to new obligations. And developing countries sought to limit talks solely to setting new targets for industrialized countries. After a long scramble for power, two working groups were set up in parallel to talk about the future. In one working group, Kyoto signatories discussed the targets for a second commitment period of the Kyoto Protocol, and the other working group, with American participation, entered into informal dialogue on long-term common measures that could be agreed upon within the UNFCCC.

Two years later, in spite of all the obstacles, a mandate for setting up a new global climate regime came out of this dialogue, the Bali

Action Plan worked out at the Bali climate conference. Its purpose is to gain the commitment of the United States and important emerging countries to set emissions reduction targets. A new de facto global agreement (or even several of them) began to take shape. The latest report from the Intergovernmental Panel on Climate Change (IPCC) had appeared a few months earlier and the panel had just been awarded the Nobel Peace Prize →p.458. The outcome of the report may have contributed to reaching an agreement. Scientists had come to the conclusion that human activity alone was responsible, with more than 90 percent probability, for most of the global warming observed in the last fifty years—in the 2001 IPCC Report, this probability had been estimated at only 66 percent.

The Bali documents are still at the heart of today's climate negotiations within the framework of the United Nations. Negotiators at the Bali conference succeeded in establishing a consensus that all industrialized countries should make "measurable, reportable, and verifiable" efforts to lower greenhouse gas emissions. Even the United States was now back in the process. It had faced considerable pressure from the plenary session. In return, developing countries promised to take appropriate action to reduce emissions according to their national abilities. Verifiability is an essential element, and the abbreviation MRV (measurable, reportable, and verifiable) is now widely used for this concept. Verifiability applies both to the emissions reduction targets of industrialized countries and to the measures taken by developing countries. It was foreseeable that it would not be easy to agree on a text for a binding resolution by 2009. Nevertheless, many were counting on a new president in Washington, D.C., and even at the 2008 Poznan Climate Change Conference, many people were still not aware that Barack Obama, at that time president-elect of the United States, would not be able to easily change power relationships in his country →p.464.

But other things had changed too. In 2006, China became the world's biggest emitter of greenhouse gases, surpassing the United States. The International Energy Agency reported that China emitted 17 percent more CO_2 from burning fossil fuels in 2008 than the U.S. did; as recently as 1990, China's emissions had amounted to less than half of the emissions released in the United States. Per capita emissions in China in 2008 were, however, only one-quarter of those in the United States, but already five times higher than in India.

Global emissions of greenhouse gases from fossil fuels continued to rise. The International Energy Agency reported an increase of a good 40 percent from 1990 to 2008. However, overall emissions in industrialized countries remained nearly constant according to reported statistics, and in industrialized countries with Kyoto targets they even decreased by 9.2 percent.[1] In an UNFCCC inventory taken in November 2010 for the Cancún climate conference, experts arrived at similar figures for industrialized countries.[2] They even calculated an overall decrease of 6.1 percent for greenhouse gas emissions for all industrialized countries, including the United States, in the same time period, without taking into account sinks like forests (LULUCF). This decrease must be attributed primarily to Russia and former Eastern Bloc countries. Their emissions dropped by 36.8 percent as a result of the economic breakdown after the fall of the Soviet Union, while emissions in all other industrialized countries rose altogether by 7.9 percent.

The political environment also changed. Early in 2006, George Bush initiated the Asia Pacific Partnership on Clean Development and Climate, and before the conference in Bali he also called together a forum of the largest CO_2-emitting countries. This forum, which has continued during the Obama Administration under the name of Major Economies Forum on Energy and Climate, counts as its members more than a dozen countries that are together responsible for 80 to 90 percent of global emissions. There has been speculation that this could undermine the United Nations' leading role in global climate negotiations. In addition, the worldwide financial crisis that began in 2007 created tension everywhere in the economic system. The battle for public funding intensified, and willingness to make financial investments in a more climate-friendly future generally fell off.

The plenary of the Bali climate conference had mandated an enormous task for the following two years, in the hope of finally cutting through the Gordian knot of Kyoto. This was why the conference in Copenhagen became an oversized media event, preceded by a great deal of publicity, and built up as the kind of summit that would determine the fate of the world. But the time allotted to the task was simply too short. Nevertheless, organizers did not begin until the second half of 2009 to communicate that there was not enough time left to work out a legally binding agreement—too late.

Expectations for the UN Copenhagen Climate Change Conference beforehand were as enormous as the frustration that followed afterward. The conference registered more than 40,000 participants. Its leadership was unable to cope, and Obama's last-minute effort to word a text with powerful emerging countries, more or less in a back room, which could have been approved by the plenary session as a conference document, failed. Most countries in the UNFCCC, even the European Union, only found out about this behind-the-scenes negotiating session afterward. A great sense of annoyance over the course of action and its results prevailed, since many saw outcomes as too weak and negotiations as too opaque. After a spiteful closing debate that went on through the night until late the next morning, and after long consultations, the Copenhagen Accord was not officially adopted by the member states but only "taken note of." Above all, delegates from developing countries expressed fury over the back-room compromise that had excluded the rest of the community of states.

The document contains various promises to developing countries on technology transfer and the financing of measures to combat climate change, but also resolutions to halt the destruction of tropical forests. One appendix contains the reduction targets announced by various states, or promises of programs to decrease their greenhouse gas emissions. After the Copenhagen conference, well over eighty countries submitted relevant declarations on reducing emissions; among them were China, the United States, the EU countries, and all other major emitters. Not until the following Conference of Parties (COP16) in Cancún late in 2010, however, was it possible to officially incorporate the most important points of these documents into the Cancún Agreements as the basis of a new agreement. In contrast to the Copenhagen conference, the conference in Cancún was said to be an earnest working conference right from the start.

A truly binding climate regime, however, as was intended for Copenhagen, had to be postponed for years. But the climate conference in Durban succeeded in establishing a new Ad Hoc Working Group, the Durban Platform, with a mandate to develop a protocol or an other “agreed outcome with legal force under the Convention applicable to all Parties” to come into effect in 2020. It shall be adopted no later than 2015. What legal form it will take is still just as open as the level of ambition for the mitigation goals. But there is also a work plan launched to enhance this ambition. An expert report by the United Nations Environment Programme had already noted before the Cancún conference that the reduction goals listed in the Copenhagen Accord are far from sufficient to achieve the target set in the Cancún Agreements for limiting global warming to 2 degrees Celsius (3.6°F). Even with consistent enforcement of the steps discussed in Mexico, only about 60 percent of necessary emissions reductions will be achieved by 2020. Experts say that at least 5 billion tons of CO_2 equivalents must be additionally reduced to reach the target—this is called the five-gigaton gap.

However, should negotiators succeed by 2015 in adopting a new global, binding agreement with the capacity that is intended, and if it includes a useful interpretation of “measurable, reportable, and verifiable,” then it would regulate a quantity of emissions many times greater than does the first commitment period of the Kyoto Protocol. The fact that the large developing countries are also willing to accept the still-to-be-defined rules permitted the U.S. to support the Durban Platform. At the same time, the promise of a future agreement among all parties also motivated the EU to continue with a second commitment period of the Kyoto Protocol even though not every relevant emitter will be part of it.

1 IEA, *CO_2 Emissions from Fuel Combustion: Highlights,* OECD/IEA, Paris, 2010.
2 UNFCCC, *National Greenhouse Gas Inventory Data for the Period 1990–2008,* FCCC/SBI/2010/18, Cancún, 2010.

MONEY MAKES THE WORLD GO ROUND

Most environmental problems are a consequence of economic development. In the course of history, poor countries have usually contributed less to these problems than rich countries have, and they have less money to do something about it. This is reflected in climate policy. Money and technology transfers are supposed to create a balance between North and South.

Money plays a decisive role in international climate policy. Here we do not mean the money used in industrialized countries to promote energy efficiency and the use of renewable energy sources, such as the subsidy programs for solar and wind energy in Germany that distribute billions of euros each year. Nor is this about the money put into research and development for new technology in cars with electric or hybrid drives, for efficient housing technology and home appliances, or for "green" computers. Here we are talking about the money that flows from industrialized to developing countries.

The prosperity gap between rich and poor, between industrialized and developing countries, with the latter acting jointly in the Group of 77 and China within the framework of the United Nations, is a central feature of climate negotiations →p.464. The very different roles played by industrialized states and developing countries in creating the climate problem were already being brought up again and again by the end of the 1980s, before negotiations for a climate convention had even begun. When the Berlin Wall fell, the former Soviet Bloc broke apart, and the Cold War was over, hopes for a peace dividend prevailed. Many believed that the money that had once flowed into the military sector could now be used to resolve development and environmental problems.[1].

The epochal 1992 Earth Summit in Rio de Janeiro was officially named the United Nations Conference for Environment and Development (UNCED), suggesting a promise that the Group of 77 and China believed was never delivered. Indeed, people in industrialized states viewed it mostly as an environmental summit having very little to do with development.

The concept of finding a balance between North and South was enshrined as early as 1987 in the Montreal Protocol on Substances that Deplete the Ozone Layer. Late in 1989, the UN General Assembly explicitly acknowledged the right of developing countries to claim financial support. The mandate for the Earth Summit, then under preparation, was to find ways and means to solve environmental problems in connection with development, and to make available "new and additional financial resources, particularly to developing countries, for environmentally sound development programs and projects." The UN resolution adopted at the same time to work out the framework climate convention had similar wording.

This resolution also maintained the importance of developing ideas for making "new and additional financial resources" available to poorer countries, needed to cover the costs arising from implementation of the climate convention and relevant measures for action. The promise of such financial resources became a central demand of the Group of 77 and China when negotiations began in 1990—it was clear that without this money they did not intend to meet their obligations in climate agreements. "New and additional" were the words cited again and again when it came to finances, and they have now become a familiar incantation in international environmental policy talks. Assurances of this kind to developing countries are therefore enshrined in the UN Framework Convention on Climate Change that went into force in 1995, just as they are in the Kyoto Protocol.

In recent years, hardly a conference of parties of the climate convention has come to an end without countries of the South being made promises of "new and additional" financial resources in some form or another. Multilateral and bilateral funds, mechanisms, and cooperative programs have become so numerous that only specialists are able to keep track. The Global Environmental Facility (GEF) was the first financial institution to come into play. It was created shortly before the Earth Summit in Rio de Janeiro to finance environmental projects and programs in the developing world together with the World Bank, the UN Environment Programme (UNEP), and the UN Development Programme (UNDP). In the UNFCCC as well, it was given the assignment—at first only temporarily—to allocate relevant financial resources. But since the beginning of negotiations over the climate convention, there has been a fierce battle over who has how much to say about controlling finances and their distribution. Within this context, the GEF has been repeatedly criticized by many poorer countries for being too close to the World Bank and industrialized countries. However, the latter refuse to cede too much control over the distribution of resources to recipient countries. In order to give developing countries more of a say in management committees, the GEF has, over time, detached itself from the World Bank.

In the meantime, within the scope of the climate convention, the GEF now also heads two funds established in 2001 at the Conference of Parties (COP7) in Marrakesh: the Least Developed Countries Fund (LDCF) and the Special Climate Change Fund (SCCF). Together with the World Bank, it also supports the management of the third fund of the climate convention, the Adaptation Fund set up half a year earlier. This fund administers and distributes the money accumulated from projects in the Clean Development Mechanism of the Kyoto Protocol →p.411. It enables industrialized states to achieve some of their emissions reductions through projects in developing countries, which, however, must meet a number of standards; in return, for example, they must turn over 2 percent of the emissions reduction certificates (CER) gained to the Adaptation Fund. The GEF is today responsible for distributing financial resources within other United Nations environmental conventions as well, in total for about $500 million a year. Since pledges from the GEF often induce others to make financial contributions, this figure goes up to several billion dollars per year.

There are numerous other multilateral financing programs in place in addition to the GEF and its funds. For financing the protection of tropical forests →p.518, there are the REDD+ Partnership, the UN REDD Programme, and the Forest Carbon Partnership Facility. Other institutions include the Climate Investment Funds, the Carbon Partnership Facility of the World Bank, and regional development banks. There are also a great variety of bilateral activities. The World Bank says that the entire sum distributed through these channels in 2009 could have reached about $10 billion—far too little, as experts point out.

The Copenhagen Accord, formulated in 2009 more or less at the last minute by a few major economic powers, therefore held substantial new financial pledges. Developing countries were promised a total of $30 billion in a "fast start" for the period from 2010 to 2012 to give them quick support for mitigating and adapting to climate change; pledges from industrialized countries were submitted well before the follow-up conference in Cancún began. This funding is supposed to reach $100 billion a year starting in 2020. These contributions were also set out in the Cancún Agreements late in 2010 in connection with the new global climate change regime that is foreseen.

But no one knows whether this is really enough and whether recipient countries will be able to set up institutional processes on time that can absorb the flow of money. The Group of 77 and China is asking for 1.5 percent of the gross domestic product of industrialized countries—this corresponded to around $600 billion in 2009 in OECD countries, or nearly four and a half times the amount of total development aid from countries belonging to the OECD Development Assistance Committee. Yet individual voices in Cancún called for 6 percent and more of the gross domestic product of industrialized countries. Nicholas Stern of the London School of Economics, renowned as a climate economist since 2006, estimates that around $200 billion are needed, half for emissions reduction and half for adapting to a changed climate →p.411. The World Bank's figures are even higher by another few dozen billion dollars. Christiana Figueres, executive secretary of the climate convention, therefore describes the $100 billion set out in the Cancún Agreements as a bare "minimum," but says it is nonetheless important for this sum to be put on the table very soon.

Even these $100 billion per year pose a great challenge to the financial system. This was the verdict from the High-Level Advisory Group on Climate Change Financing, a group of economists and finance experts from North and South appointed by UN Secretary-General Ban Ki Moon shortly after the Copenhagen conference. But the group also came to the conclusion that this figure could actually be raised if a variety of state and private channels could be combined in an appropriate manner. They maintained that it was crucial for industrialized states to introduce strict standards for domestic reductions in emissions, and for the price of emitting a ton of carbon dioxide to reach at least $20 to $25 by 2020. The higher the price, the greater the reduction in emissions, and the more money to be mobilized through different measures such as a tax on emissions trading.

Similar to the situation twenty years earlier, talk about finances at the UN Climate Change Conferences in Cancún and Durban centered on control and the shape of the institutions distributing financial resources. How would financial flows be structured, and who would sit on what committee? Nevertheless, the participants at the Durban conference did succeed in making the Green Climate Fund operational. The fund established in Cancún to distribute resources is governed by an administrative board of twenty-four members, half of them appointed from industrialized countries, the other half from developing countries. It will have, among other things, a special facility to promote the participation of the private sector in financing—a decision that has been criticized by both developing countries and non-governmental organizations. But the main problem will be to get sufficient funds. Therefore a special work program has been set up in Durban to mobilize a wide range of financial resources. The fund's finances are to be administered by the World Bank, at least for the first three years.

The Green Climate Fund's success will greatly depend on credibility and transparency in its dealings with flows of money. Credibility can be improved if the money is used "wisely" in the next ten years, say Ban Ki Moon's financial experts. A high degree of transparency both in the availability and the use of financial resources could allow money to flow more easily. The issue of transparency also needs a clear answer to the question of which money really is "new and additional"—whether it is something more than just old wine being sold in new bottles. Critics have pointed out that even Germany's promises include financial commitments that have been made before. Added to that, not all participants understand the same thing under "new and additional."

However, even more basic misgivings may need to be dealt with. There is genuine concern that the dynamic carbon market aimed for here, whose incentives are bound to attract a large number of private investors, could develop its own dynamic and once again endanger underlying environmental goals and threaten development opportunities for poor populations.

1 UNCED, AGENDA 21, Paragraph 33, 16e, Rio de Janeiro, 1992.

THE POINT OF VIEW FROM INDIA: EQUITY, THE NEXT FRONTIER IN CLIMATE TALKS

Sunita Narain, director of the Indian environmental organization the Centre for Science and Environment (CSE), and editor of *Down To Earth,* a renowned science and environment magazine, commented in late 2011 on decisions made in Durban from the perspective of developing countries.

In 1992, when the world met to discuss an agreement on climate change, equity was a simple concept: sharing the global commons—the atmosphere in this case—equally among all. It did not provoke much anxiety, for there were no real claimants. However, this does not mean the concept was readily accepted. A small group of industrialized countries had burnt fossil fuels for a hundred years and built up enormous wealth. This club had to decide what to do to cut emissions, and it claimed all countries were equally responsible for the problem. In 1991, just as the climate convention was being finalized, a report, released by an influential Washington think tank, broke the news that its analysis showed India, China, and other developing countries were equally responsible for greenhouse gases. Anil Agarwal and I rebutted this and brought in the issue of equitable access to the global commons. We also showed, beyond doubt, that the industrialized countries were singularly responsible for the increased greenhouse gases.

In 1992, it was accepted that the occupied atmospheric space would need to be vacated to make room for the emerging world to grow because emissions are an outcome of economic growth. This acceptance recognized the principle of common but differentiated responsibilities in reducing emissions. A firewall was built to separate those countries that had to reduce emissions to make space for the rest of the world to grow. That year, in Rio de Janeiro, the world was talking about drastic cuts of 20 percent below the 1990 levels to provide for growth as well as climate security. Even in that age of innocence, the negotiations were difficult and nasty. The United States argued that its lifestyle was non-negotiable and refused to accept any agreement specifying deep reductions. In 1998, the Kyoto Protocol set the first legal target for these countries much below what the world knew it needed to be.

Two decades later, the idea of equity has become an even more inconvenient truth. By now there are more claimants for atmospheric space. Emerging countries have emerged. China, which in 1990, with over a quarter of the world's population, was responsible for only 10 percent of annual emissions, contributed 27 percent by 2010. So, the fight over atmospheric space is now real. While the rich countries have not reduced emissions, the new growth countries have started emitting more. In 1990, the industrialized countries accounted for 70 percent of the global annual emissions. In 2010, they accounted for 43 percent, but this is not because they have vacated space. The new growth countries — China in particular — have only occupied what was available. Emission reductions proposed twenty years ago have still not been committed or adhered to. In fact, in most already industrialized countries emissions have either stabilized or increased. In coal and extractive economies, like Canada and Australia, emissions have risen by 20 percent and 46 percent, respectively.

The world has run out of atmospheric space and certainly of time. Will the rich, who contributed to emissions in the past and still take up an unfair share of this space based on their populations, reduce emissions? Or will the emerging countries be told to take over the burden? This is the big question, and an inconvenient one at that.

And mind you, climate change is not the problem of the present but past contributions. The stock of greenhouse gases in the atmosphere has a long life. This means that any discussion on how the carbon cake will be divided must take into account those gases emitted in the past and still present. So while China accounts for 27 percent of the annual emissions, in cumulative terms (since 1950) it still accounts for only 11 percent. Similarly, India contributes 6 percent to the annual global emissions, but is only responsible for 3 percent of the stock. The rich countries, with less than a quarter of the world's population, are responsible for some 70 percent of this

India is no longer just a developing country—it does what it can to be better heard at international climate negotiations. *Christopher Anderson/Magnum Photos*

historical burden. This stock of gases is responsible for an average global temperature rise of 0.8 degrees Celsius (1.4°F) and another 0.8 degrees Celsius in the future, which is inevitable. To keep temperature rise below 2 degrees Celsius (3.6°F), the world needs to cut emissions by 50 to 80 percent below the 2000 levels by 2050. Now equity is no longer a moral idea, but a tough challenge. It is for this reason that global climate negotiations reached their nadir in Durban. It is for this reason that the U.S. and its coalition are hell bent on erasing any mention of historical emissions from all texts. It is for this reason that the rich world is pointing to the emission growth in China and India, and dismissing their need for development as their obdurate right to pollute.

It is also an idea that is difficult to sell in a world distrustful of idealism and any talk of distributive justice. Even climate change negotiators do not really believe this form of climate socialism can happen. They will tell you that the world is never going to give up space, that the world is too mean to give money or technology to poor nations for transition to low-carbon growth.

But this is because they forget that climate change is the market's biggest failure. We cannot use the market for its repair. To avoid catastrophic changes it is essential to reach a collaborative agreement that will be effective. And cooperation is not possible without fairness and equity. This is the prerequisite. Take it because we must.

THE POINT OF VIEW FROM CHINA: TIME IS OF THE ESSENCE

At the end of the Cancún conference, Li Xing, deputy editor of the *China Daily,* China's official state organ, gave a summary of the position of her country on climate policy. The English-language newspaper is considered a mouthpiece of the Chinese leadership.

Cancún, December 12, 2010.... Each of us—officials, activists, and journalists—left a substantial carbon footprint just to get here. According to the host's website, I will have created 14.5 tons of carbon dioxide traveling to Cancún from Beijing and riding buses every day between my hotel and the conference venues.

All of us are working to rally the political will of more than 190 countries to ensure a better future for our children and grandchildren. Climate change is real and time is of the essence. Nations must not wait until the negotiators hammer out the legal documents to take action.

China is not waiting. It is quickening its steps to cut fossil fuel consumption, plant more trees, and above all reduce the intensity of carbon emissions per unit of its GDP, as Xie Zhenhua, head of the Chinese delegation, has repeatedly explained. Xie has also said that China is considering putting climate legislation in its Twelfth Five-Year Plan (2011–15), in addition to measures promoting energy conservation, renewable energy, and environmental protection.

In November, six Chinese government agencies announced that they were finalizing a set of national energy-efficiency regulations. Under the new rules, power companies must use at least 0.3 percent of their electricity revenues to develop programs to help factories, businesses, and households invest in energy efficiency. The companies will also be asked to save energy, with binding targets set at 0.3 percent of the previous year's maximum load. According to calculations by the U.S. NGO the Natural Resources Defense Council (NRDC), a 0.3 reduction in power usage a year

China battles climate change on many fronts—hay is laid out to stabilize dunes and put a halt to the expanding Maowusu Desert in Ningxia Hui Autonomous Region. *Michael Reynolds/Keystone*

"would roughly translate to 11 billion kilowatts per hour (kWh)—enough electricity to supply 1 million average U.S. households (or 10 million Chinese homes) for a year."

Chinese businesses are not waiting. In Cancún, their representatives have demonstrated a series of endeavors to develop low-carbon construction, green transportation, and other energy savings.

Chinese NGOs and international organizations working in China are not waiting. The Beijing-based Shanshui Conservation Center, headed by Lu Zhi, has been piloting a "panda carbon storage" program in southwest China. Under the program, Lu and her colleagues work with people in the mountainous communities to restore and manage forests with credits they've obtained from the international carbon markets. "We make sure that local communities are masters of these forests and benefit from reforestation," Lu said.

The NRDC has worked with Jiangsu province to limit the increase in its electricity generation over the past three years. By saving 3.5 billion kilowatts per hour a year, the province expects to reduce its CO_2 emissions by about 3.4 million tons. Meanwhile, the Climate

Group is working in Chinese cities to blaze a trail for low-carbon development.

It is tough to give up the road of high energy and resource consumption, yet we must change gears. "We can no longer afford to rely on excessive fossil fuel and intense energy consumption to drive our development," Xie said. Although China is still a developing country, it will contribute its share to slowing down global warming and reducing the effects of climate change.

We must take seriously the new data scientists have shared with us here in Cancún. The World Meteorology Organization reported that 2010 has been among the three warmest years since 1850. Meanwhile, glaciers from the Himalayas to Alaska to the southern-most portion of South America are receding, and the amount of ice available to reflect sunshine and keep our earth cool is shrinking, according to a report by the United Nations Environment Programme. The extreme weather around the world will only get worse, climate experts warn, if we don't begin drastic coordinated global actions to reduce greenhouse gas emissions and cut our reliance on fossil fuels. We must stop dragging our feet.

THE ROLE OF FORESTS IN CLIMATE POLICY

Global climate policy is the art of creating intelligent and foresighted rules for action in a domain that is highly interlinked but not fully understood by scientists. The inclusion of forests in climate policy shows how difficult this task is, but at the same time demonstrates that political decisions can indeed meet complex challenges. Some decisions are still pending and proof that others have worked is elusive, but the way tropical forests are protected might eventually turn out to be a success story.

The climate system and its feedback mechanisms continue to be understood in more and more detail. Our growing knowledge of these processes is also reflected in climate policy. If various relationships in nature and society are better understood, then this information feeds back into the contractual framework set up to govern this area. We can see a good example of this in the regulations that concern forests. The biosphere, and the forests belonging to it, play a central role in the natural CO_2 cycle. Terrestrial ecosystems are estimated to remove a total of 440 billion tons of CO_2 (corresponding to 120 billion tons of carbon) from the atmosphere each year in the process of photosynthesis, and approximately half of this is immediately released back again by plant respiration. A similarly large share is also rereleased to the atmosphere when animals breathe and when dead vegetable and animal material decomposes; in the medium term, CO_2 is also released through fires. A small share remains stored in soil as organic material over the long term. This continual back and forth between biosphere and atmosphere shapes the well-known CO_2 curve of Mauna Loa in Hawaii, where atmospheric CO_2 levels have been measured directly since 1958 →p.147. The concentration of CO_2 has been steadily rising for years, but the curve shows a wavy course.

This seasonal up and down of CO_2 levels is attributed to the activity of terrestrial ecosystems. Concentrations reach a seasonal high in May, while the lowest values are measured in September and October. Biomass on land, and especially the deciduous forests that are widespread in the Northern Hemisphere, use warmth and sunlight in the summer to grow, removing CO_2 from the atmosphere through photosynthesis. In the autumn, leaves fall to the ground where they decompose during winter and spring and release CO_2 together with breathing animals. Consequently, CO_2 concentrations in the atmosphere once again increase as late spring approaches. The medium and long-term development of the curve as a whole was long seen as a consequence of the release of carbon dioxide alone from fossil fuels and the cement industry.

The first report of the Intergovernmental Panel on Climate Change (IPCC) in 1990 asserted that in an "undisturbed world" the yearly uptake of CO_2 by vegetation and the release of CO_2 through the breathing of bacteria and microorganisms was more or less in balance. But in the meantime it has become clear that there are imbalances in this process that have considerable impact on the flows of carbon relevant to the climate debate. The amounts converted by terrestrial biomass are enormous compared with the amounts released by human beings due to fossil fuel combustion and cement production each year, which in 2009 came to just under 31 billion tons of CO_2. Added to that, little was known until ten years ago about how the CO_2 balance of the planet's ecosystems reacted to changing climate conditions.

Climate convention conferences, in contrast, talked about the influence of forests very early on, especially about deforestation and afforestation. During preparatory negotiations to the UNFCCC in 1991 and 1992, the United States, much to the displeasure of developing countries, was already arguing for including forests in calculations. But taking account of sinks with their potential for removing significant amounts of CO_2 from the atmosphere was contested. This was not only because many developing countries with large forested areas feared they would come under pressure, but also because there were gaps in knowledge and it was difficult to assess the effect of sinks. Nonetheless, the convention did prescribe that the sustainable management as well as the maintenance and improvement of sinks and potential reservoirs should be fostered, "including biomass, forests and oceans, as well as other terrestrial, coastal, and marine ecosystems." These were also to be taken into account in national greenhouse gas inventories.

Forests and terrestrial sinks in general did not become an explosive political issue until 1997 in Kyoto, during negotiations over the protocol. There, sinks turned out to be a central element in the tug of war between the United States on one side and the European Union and many environmental activists on the other side. American negotiators did not want to accept drastic reduction targets for emissions unless forests were taken into account. The issue became famous under the abbreviation LULUCF (land use, land-use change, and forestry). The incorporation of LULUCF was one of the bitter pills that the EU and environmental organizations had to swallow in Kyoto in order to see through the quantitative reduction targets they favored. Nevertheless, for the first commitment period the compulsory inclusion of sinks was limited to afforestation, reforestation, and deforestation as direct human interventions →p.464.

Before Kyoto, the political weight of LULUCF had obviously been underestimated by many, even though there were intense disputes over the issue in the weeks before the conference, with sinks among the most contentious points of conflict. In spite of years of talks, scientific studies that could have supported relevant decisions were still almost totally absent in 1997. "Science has given relatively little attention to biological carbon sinks until now, since they are limited in the long term. Overnight they are now moving into the center of the climate policy debate," wrote two Swiss scientists, experts on forests and terrestrial ecosystems, before the 2000 climate conference in The Hague.[1]

Soon after the Kyoto conference, the IPCC was given the assignment of researching the issue. Its *Special Report on Land Use, Land-Use Change, and Forestry,* nearly 400 pages long, appeared a good two years later, and nevertheless indicated that many questions were still unanswered. It became clear that terrestrial biological carbon flows were significantly less balanced than had been thought. In the 1990s, according to data in the report, natural processes—especially in forests in the Northern Hemisphere—probably had removed more than three times as much carbon dioxide from the atmosphere as in the preceding decade, whereby figures showed a considerable range of uncertainty. Studies indicated that annual averages had increased from 0.7 to 2.6 billion tons of CO_2 in recent decades. Current uptake is estimated at nearly 3.7 billion tons of CO_2 per year. This finding confirmed earlier studies, among them studies of American forests, which were already exhibiting this trend early in the nineties.

The report also threw light on the complex processes that play a role in the carbon balance in forests and in soils in general, as well as in savannahs or regions used for agriculture. The characteristics of soil governing its ability to store CO_2 can change within the space of a few hundred meters. Factors such as sunlight, climate, nutrient supply, moisture, and soil properties often show major fluctuations over time. The effect that a forest has on the CO_2 balance does not depend on the number and size of trees alone, but also on the species of trees, their age, what type of soil they grow in, and the impact of the microclimate that develops as leaves and dead branches fall and decompose. This complexity means that drawing up an inventory for the carbon stored in soil and in plants is extremely challenging. However, such inventories are of great significance, because soil, on global average, contains approximately four times as much carbon as aboveground vegetation does. This ratio does vary widely, however, depending on the ecosystem in question.

If forests and soil are to be calculated as sinks that remove atmospheric CO_2, then the lasting nature of this effort, called "permanence" in this context, is also important. The growth cycles of forests stretch over decades and centuries, and the carbon content of soils builds up even more slowly. In contrast, stored greenhouse gases can be released very quickly. A single major forest fire emits enormous amounts of CO_2. Peat soils that have formed over millennia release their bound carbon within a few decades, if not years, if they dry out. Caring for forests and other sinks is a long-term business. The permanence of sink contributions depends therefore not only on sustained political will and such things as putting a stop to illegal logging. Nature itself can annihilate efforts with a single blow. Even the question of how forest losses through storm damage are to be calculated must be addressed.

To set out the legal framework conditions for including forests and other sinks in climate agreements, it was important first to gather current expertise, process it credibly and transparently, and make it comprehensible. Negotiators then had to take differing interests into consideration in a way acceptable to all parties. Even in industrialized countries—and they are the ones affected by LULUCF—forests and their management do not always have the same status. In earlier centuries, forests in many countries were cleared over vast areas, much like the situation today in developing countries, until bitter experience showed the need to protect them, and reforestation was tackled. To a certain extent, these reforested areas have now become large sinks. Some states have hardly any forests, or their stands of wood are so old that they no longer bind additional CO_2. After years of negotiation, basic agreements were finally made at the Conference of Parties (COP7) in Marrakesh in 2001 on how industrialized states could calculate emissions and the sink performance of LULUCF within the framework of the Kyoto Protocol. These regulations were later worked out in more detail.

The main concern was that practices for including forests as sinks would be too loosely interpreted and would thereby completely undermine the reduction targets of the Kyoto Protocol. The full potential of biological sinks actually proved to be many times greater than the nearly 2.6 billion tons of CO_2 by which all industrialized states together had to reduce their emissions, compared to 1990 levels, during the first commitment period. As a result, a large number of regulations and definitions now have to be taken into account for calculating LULUCF. Signatory states also have to keep LULUCF contributions separate in their national inventories. Another area of conflict emerged around the question of the maximum number of removal units (RMUs) individual countries could be credited for as sink contributions from the care of forests, in addition to straightforward afforestation. In the end, agreement was reached that each industrialized state had to declare in advance which terrestrial sinks in agriculture and forestry it wanted to register, and to what extent. And a limit was placed on the maximum contribution that could be calculated for forest management. For that purpose, each country was assigned an upper limit for the first commitment period from 2008 to 2012 to restrict the declaration of unjustifiable contributions of natural sinks as a measure to mitigate climate change.

The inclusion of forest management projects in the Clean Development Mechanism (CDM) was an even more controversial issue

“My father rode a camel.
I drive a car.
My son flies a jet airplane.
His son will ride a camel.”

during negotiations. With this mechanism, the Kyoto Protocol enabled industrialized states to add certain reductions of greenhouse gases in developing countries to their own accounts, taking advantage of the fact that developing countries did not have to meet any reduction targets themselves →p.401. The EU feared that a generous ruling on forest projects within the framework of the CDM could serve industrialized countries as a loophole for meeting their own reduction targets. As an outcome of this concern, it was agreed that an industrialized country must guarantee that a reforestation project it sponsored in a developing country actually created new forest that would not have been planted without this financial support. But how can it be decided whether such a project would have happened without having been sponsored as a CDM project? Likewise—and especially important for this issue—only a project with permanence counts.

More details of the Kyoto Protocol worked out in later negotiations allowed afforestation and reforestation projects to be part of the Clean Development Mechanism. But this option is actually rarely used. This may be due to the great complexity of such undertakings and the many conditions attached to them. For instance, only time-limited emission reduction certificates can be procured with CDM forestry projects. After a project runs out, at the latest after sixty years—or even earlier, if a fire or an infestation of beetles destroys the forest—these certificates must be replaced by other credits. After all, less than 1 percent of the CDM projects registered in 2010 were devoted to afforestation or reforestation.

Not until the 2007 climate conference in Bali were forests in developing countries—mostly tropical rainforests—prominently featured on the agenda again. But now, five years before the first commitment period of the Kyoto Protocol ran out at the end of 2012, it was a matter of setting guidelines for the future. Ever since the Bali Climate Change Conference, the focus has been on setting up an effective comprehensive and global climate change regime, in addition to the Kyoto Protocol, to significantly reduce greenhouse gas emissions in coming years and decades →p.500. An important contribution to this goal is the protection of tropical rainforests. Since the Bali conference, the abbreviation REDD+ has stood for Reducing Emissions from Deforestation in Developing Countries, but this is often written out today as Reducing Emissions from Deforestation and Forest Degradation. The plus sign is understood in general to indicate that other activities may be included as well, such as the protection of forests that hardly change anymore, sustainable forestry, or a targeted increase in carbon storage.

Putting a halt to deforestation in the tropics became of political interest because clearcutting contributes a large share of the annual CO_2 emissions generated by human activity—during the 1990s, this came to about 20 percent. In the run-up to the Bali conference, it became clear that logging could be prevented at a relatively low cost. A year earlier, economist Nicholas Stern, at the request of the British government, had taken a detailed look at the economic aspects of climate change. *The Stern Review on the Economics of Climate Change* found that 70 percent of the CO_2 emissions released by clearcutting—4.9 billion tons yearly—were generated in eight countries. His report estimated the cost of preventing these emissions at "only" $5 to $10 billion a year—or an average of $1 to $2 per ton of CO_2. This is a fraction of the $10 to $20 paid in the years 2010–11 for each ton of CO_2 equivalents not emitted. The countries in question—Bolivia, Brazil, Cameroon, the Democratic Republic of Congo, Ghana, Indonesia, Malaysia, and Papua New Guinea—all lie in the tropics. The idea is that industrialized countries pay developing countries to maintain their forests and manage them sustainably.

Since then, REDD+ has been an important point on the Bali Road Map, the framework for the new global climate change regime, whose adoption was originally announced for 2009 in Copenhagen—and which in December 2011 in Durban was postponed until 2015. Nonetheless, the conferences in Cancún (2010) and Durban succeeded in setting up important cornerstones for the framework of REDD+.

It became apparent that REDD+ should be a relatively easy problem to solve within the comprehensive negotiation package for the new climate regime. However, many countries would be affected here as well in very different ways, including in the South. And many questions that emerged during the process of working out how to calculate sinks within the framework of the Kyoto Protocol (LULUCF), and during negotiations on the Clean Development Mechanism in relation to forests, still need to be settled in a compatible way. For instance, it is important to make sure that counterproductive land conversion is not fostered. This can involve logging to clear land for large palm oil plantations that produce biofuels. There are examples of projects like this in Southeast Asia, which have driven out local indigenous populations, and reduced the diversity of species and the long-term contribution of sinks for CO_2. Likewise, it is important to prevent situations in which REDD+ projects sponsored by industrialized countries are misused as cheap solutions to meet their own reduction targets.

While negotiations continued, several industrialized states, such as Norway and France, began committing themselves bilaterally to protecting tropical rainforests. A REDD+ partnership was founded in May 2010 in Oslo that made financial pledges of more than $4 billion. By the end of the year, over seventy countries had joined the partnership, among them Switzerland and Germany. With hundreds of voluntary projects, the partnership is looking to develop the methods and institutions needed to protect tropical rainforests. Norway concluded a contract for billions with Indonesia, which encompasses a comprehensive strategy in three phases. Starting in 2014, Indonesia will receive money for verifiable emissions reductions. Key concepts in such projects include the creation of relevant know-how in developing countries, the guarantee that measures will be long-lasting and sustainable, the protection of biodiversity, and the fair inclusion of indigenous populations and the protection of their rights. There is a great deal of debate on the "measuring, reporting, and verifying" of success. It is also urgent to rule out corruption. Experience shows that it can undermine all efforts.

Since the Cancún Agreements, developing countries are now officially called upon to contribute to lowering greenhouse gas emissions through measures undertaken in the forestry sector. National strategies are supposed to be designed to this end, and a transparent system must be installed to observe the development of forests and to report on their condition. In specific terms of the Durban decisions, REDD+ asks for slowing, halting, and reversing forest cover and carbon loss. In turn, industrialized countries are called upon to support developing countries in these efforts, in particular with technology and financing. But even after the 2011 Durban conference details remained to be resolved.

Delegates in Cancún had also drawn up a long list of criteria that programs and projects must take into account. Important ones among them are the preservation of environmental integrity, the fair participation of population groups affected by such programs and projects, respect for the sovereignty of developing countries, and precautionary measures to prevent logging from being shifted to other areas. The question about the extent to which REDD+ projects should be incorporated into the carbon market remained to be clarified. But a year later the Durban conference moderated the language, what critical voices denounced as weakening the safeguards. The sources of financing also, however, remain controversial. The Durban text accepts a "wide variety of sources, public and private."

Organizations of indigenous peoples, however, have little sympathy for all these plans. They are intensely annoyed not only that relevant documents fail to call for the explicit agreement of the indigenous groups affected. They also criticize REDD+ and the carbon market in general as the wrong way to continue. They believe that industrialized states are simply passing their burden on to poor countries.

1 Andreas Fischlin and Jürg Fuhrer, "Die Klimapolitik bringt die Wissenschaft an ihre Grenzen—Die Herausforderung des Kyoto-Protokolls für die Ökologie" [Climate policy pushes science to its limits—The challenge of the Kyoto Protocol for ecology], *Neue Zürcher Zeitung,* November 9, 2000.

pp. 528–29 Evergreen tropical rainforest. Bali, Indonesia. *Remi Benali/Corbis*

p. 530 Tropical cloud forest, 1,950 meters above sea level. Braulio Carrillo National Park, Costa Rica. *Gregory G. Dimijian/Keystone*

p. 531 Forest of coast redwoods, Redwood National Park, California, USA. *Stan Osolinski/Oxford Scientific/Keystone*

pp. 532–33 Clearcutting in Kalimantan, Indonesia. *Luca Tettoni/Corbis*

p. 534 Eucalyptus grove in a temperate climate. Madeira, Portugal. *A. Hahner/EPA/Keystone*

p. 535 Old oak forest in the New Forest in the southeast of England. *Terry Heathcote/Oxford Scientific/Keystone*

p. 536 A boreal coniferous forest burns in northern Russia. *Tanya Makeyeva/AP/Keystone*
Burnt forest area in the south of France. *M. Luccioni/Keystone*

p. 537 Deforestation and erosion in Haiti. *Olivier Coret/Corbis*
Bark beetle infestation in a boreal coniferous forest. British Columbia, Canada. *Taylor S. Kennedy/Getty Images*

p. 538 A bayou with cypresses near Gibson, Louisiana, USA. *Ian Berry/Magnum Photos*

p. 539 Boreal coniferous forest at Reflection Lake, Mount Rainier National Park, Washington, USA. *Paul Edmondson/Corbis*

p. 541 Deciduous tropical rainforest in the Yungas jungle, Argentine Andes. *Philippe Psaila/Science Photo Library/Keystone*

pp. 542–43 Mixed forest in Vermont, USA. *Pete Davis*

p. 545 Cloud forest in Southwest National Park in Tasmania, Australia. *Ted Mead/Photo Library/Keystone*

"At first I thought I was fighting to save the rubber trees, then I thought I was fighting to save the Amazon rainforest. Now I realize I am fighting for humanity."

 Chico Mendes (1944–1988), Brazilian rubber tapper, trade union leader, and environmental activist

WHO
WILL DEC
THE FUT

CIDE
URE?

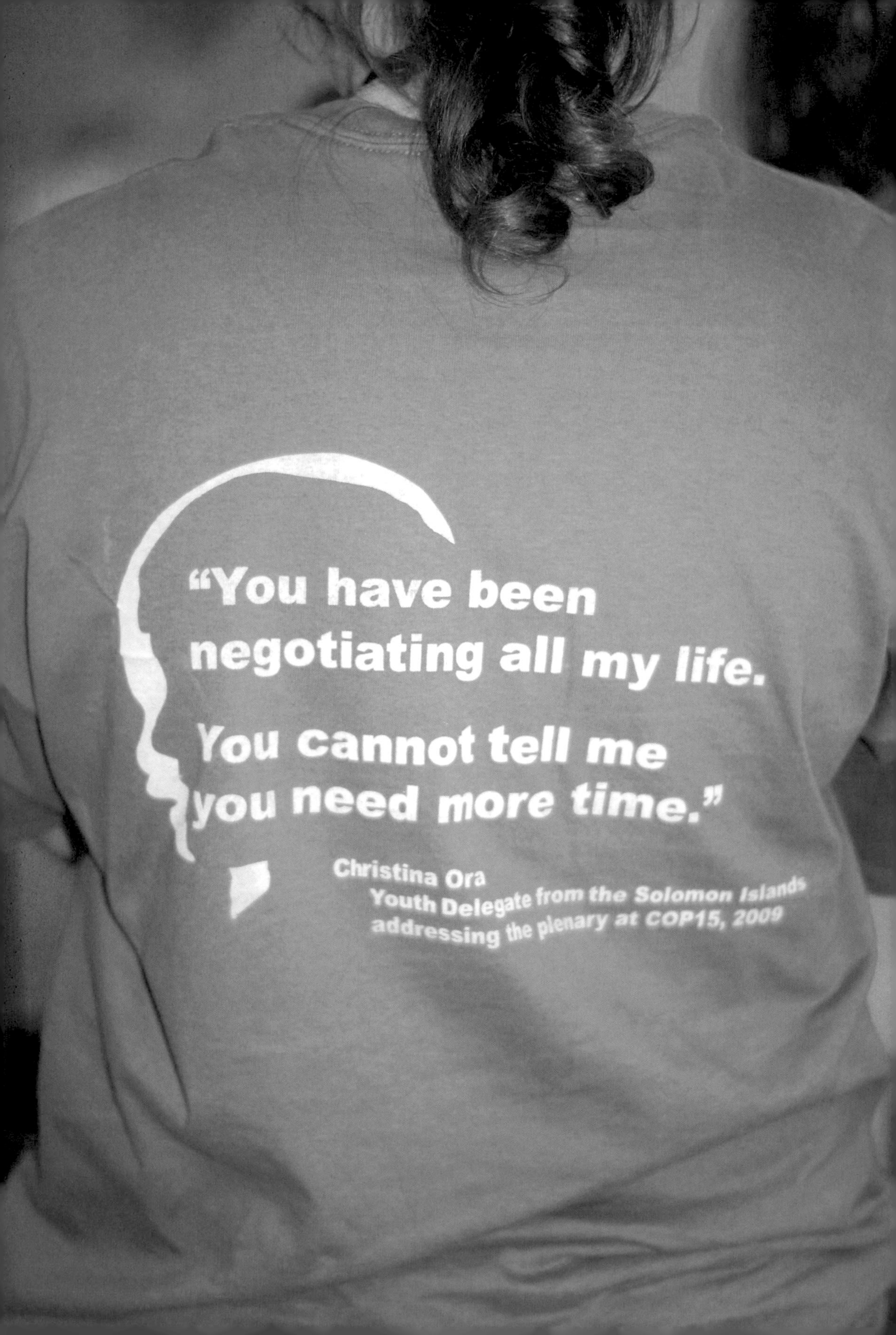

"You have been negotiating all my life. You cannot tell me you need more time."

Christina Ora
Youth Delegate from the Solomon Islands
addressing the plenary at COP15, 2009

It will not help to turn a blind eye—climate change is already in full swing. Even though meteorologists are right to point out time and again that natural disasters in recent years are nothing more than extreme weather events, the accumulation of record storms and once-in-a-century catastrophes, of floods, blizzards, and heat waves, and of forest fires, hurricanes, landslides, and year-long droughts are largely what climate scientists now expect as a consequence of global warming.

We will have to live with climate change, and so will future generations. Even if we succeed in quickly and effectively curbing the increase in anthropogenic greenhouse gas emissions, or indeed reducing them, the effects of climate change will still be dire. If our efforts do not take hold, it is highly probable that global warming will lead to unpredictable and irreversible damage within a few decades.

Climate scientists are careful not to make their forecasts sound so dramatic. Their matter-of-fact research and analyses give us figures and data that only describe the risks we run if we do this or do that. But we do not need the gift of prophecy to envisage that a large share of humankind after us will have to live in a world that we would not want to live in ourselves.

Nevertheless, pollsters have noted that environmental and climate issues have in recent years lost their urgency for the public, although a study by the European Commission indicates that about 50 percent of EU citizens think climate change is a serious problem. The international financial crisis has recently been given almost exclusive media coverage, dominating public attention and further weakening the momentum of climate politics. If people are asked about their most acute worries, they rank unemployment, the economic crisis, crime, immigration, and other issues far above environment and climate change.

There are several explanations for this paradoxical discrepancy, the simplest of which we all know from personal experience—it is natural to feel more threatened by acute problems than by events that will occur in fifty or a hundred years. Even though it may be against our better judgment, we are all too happy to be placated by statements that perhaps the looming disaster will not happen, or that human ingenuity will conjure up solutions on time and there is no need to act now.

The fact that climate processes are difficult to understand helps to suppress fears. In the end, the layperson has no choice but to believe scientists, or not. Stakeholders who expect disadvantages from mitigating climate change find it easy to team up with spin doctors to downplay the warnings of climate scientists, and to translate, for example, individual errors and uncertainties in the *IPCC Fourth Assessment Report 2007* into general skepticism about climate research.

Climate policy is not a rewarding business for politicians—commitment to effective measures is not necessarily popular. Politicians prefer tackling concerns that promise tangible benefits immediately, and that will incur costs only when they have left office. But climate policy calls for just the opposite. It generates high costs today, while benefits will not become visible for decades, or even later.

This applies at the national, but even more so at the international level. For politicians, representing their country's interests first means protecting domestic industry against international competition. Lately, as major emerging countries have started using their growing economic power to vigorously demand their proper place in the global economy and a fair share of the wealth of the world for their huge populations, this global power struggle has dominated almost all areas of international politics. Climate policy is often but a pawn in the struggle for economic supremacy between Western powers like the United States and the leading newly emerging economies, China, India, and Brazil.

Global climate policy is, however, hardly compatible with individual national interests, simply because the climate does not stop at national borders. Without consensual agreement, offering emerging markets and developing countries the same opportunities to enjoy prosperity, there will only be losers in the long run. The conflict at issue is easy to identify but difficult to resolve. On one hand, there is no real argument why an average American should produce twice the quantity of emissions as a European, five times as much as a Chinese, and twenty times as much as an African. On the other hand, goals for mitigating climate change will not be achieved if the most populous developing countries even merely double their current energy consumption.

This raises a number of novel questions. Can we, as inhabitants of industrialized countries, reduce energy consumption enough for emerging and developing countries to use more? How much prosperity do we have to forego so that other countries can increase their own? Can solar and wind energy replace fossil fuels enough to allow continued growth of the global economy? Or do we, in addition to saving energy, need to generally reduce the consumption of resources and raw materials to protect the climate and the environment, as more and more scientists think we must?

The concept of the ecological footprint can serve as a benchmark for answering these questions. It is a way of comparing our actual consumption of natural resources with what nature is able to supply and cope with in the long run. Other evaluation methods are based on CO_2 emissions (CO_2 footprint) or on overall energy consumption

(2,000-watt society). No matter which concept is used, each one shows that the populations of Western industrialized nations are living far beyond their means.

On the other hand, various studies, for instance by the Stiglitz Commission set up by the French government and headed by Nobel Prize-winning economists Joseph E. Stiglitz and Amartya Sen, or by the Novatlantis group in the ETH domain in Switzerland, suggest that a sustainable or "green" economy would be possible, and that even in the wealthiest parts of the world no one would have to relinquish quality of life in an unacceptable way. In such an economy, prosperity would not be defined only in terms of gross domestic product, but would encompass the idea that quality of life includes aspects beyond material factors.

Our current concept of wealth is based on illusions, on the central myth of the industrial age, which is that a better life can be had only through an ever-increasing exploitation of nature. It also relies on the myth that nature can be mastered by technology, and that the menace of natural disasters can be minimized to a "reasonable" level through ingenious technical invention.

We could have known for quite a while that something is wrong with these dreams. Members of the Club of Rome in 1972 foresaw this in their groundbreaking report, *The Limits to Growth,* as did the editors of *The Ecologist* magazine in their book, *A Blueprint for Survival.* At the latest, this illusion was shattered by the triple environmental disaster that occurred on March 11, 2011 in Japan. Even if the earthquake, the tsunami, and the meltdowns at the Fukushima nuclear power plant are not climate-related disasters, they prove with brutal clarity that humankind has as little control over natural forces as it has over its own technologies, and that despite all the differences between the kind of danger posed by earthquakes, nuclear technology,

or climate disasters, nothing can protect us from incalculable damage. Nonetheless, in the case of climate policy, it is now in our hands to exclude or at least limit these risks. It just needs to be done.

There are no more excuses. Each individual will have to decide whether they want to change their behavior or whether they don't care what happens to future generations. We cannot count on politics or the economic system to lead the way by their good example. Both forces will follow when we change our attitudes and behavior as citizens, voters, and consumers and insist on what we want. In spite of all the difficulties, it is only then that negotiators at climate conferences will no longer take decades to finally agree on effective measures, and put a halt to global warming.

René Schwarzenbach
Lars Müller
Christian Rentsch
Klaus Lanz

p. 548 The Secretariat of the Pacific Regional Environment Programme (SPREP), www.sprep.org
pp. 554–55 *Oliver Brenneisen/Bilderberg*

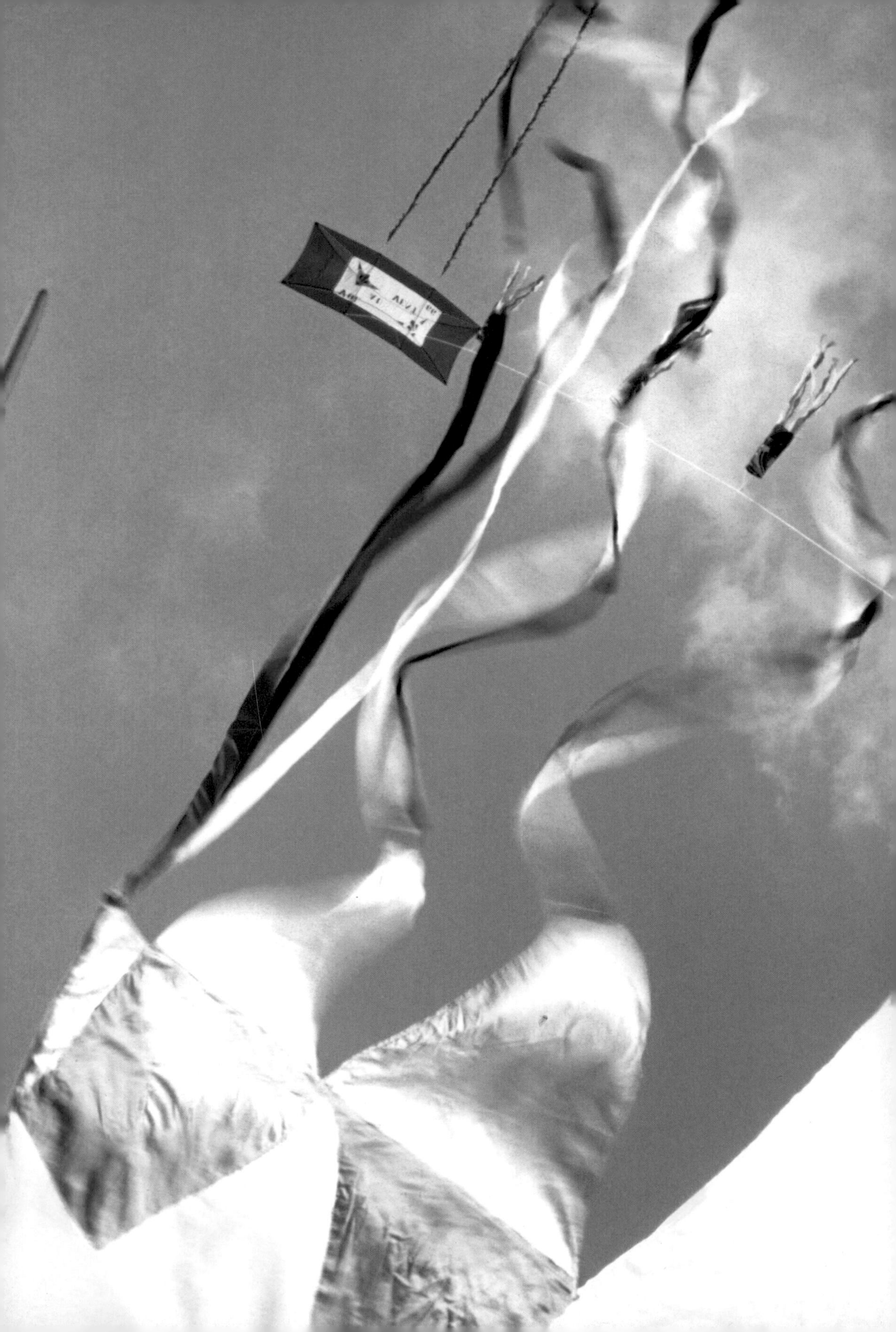

“Our climate will never be the same again. We have to understand the climate of the past to know what it will be like in the future.”

Stefan Brönnimann, climate scientist, University of Bern

“Once glaciers are gone, deleting the word ‘Alpenfirn’ from the Swiss national anthem will be the least of our problems—and not only Switzerland, but the rest of the world too, will appear very, very unfamiliar to us!”

Andreas Fischlin, systems ecologist, Swiss Federal Institute of Technology Zurich

“Whoever thinks building new gas- and coal-fired power plants is the answer to Fukushima is just replacing one evil with another.”

Nicolas Gruber, environmental physicist, Swiss Federal Institute of Technology Zurich

“The climate problem is not like a meteorite hurtling toward us while we watch helplessly, and it’s not like we unwittingly did something wrong a long time ago. We are causing climate change, here and now, every day, and we keep on doing it with complete determination.”

Reto Knutti, climate physicist, Swiss Federal Institute of Technology Zurich

“Our wasteful lifestyle is an important if unintentional contributor to climate change. The idea that we are entitled to this lifestyle doesn’t help to make it easier to accept effective climate policies.”

Gertrude Hirsch Hadorn, philosopher, Swiss Federal Institute of Technology Zurich

“Immediate and substantial cuts in greenhouse gas emissions are needed to effectively deal with climate change—and that calls for a profound economic paradigm shift toward creating a low-carbon society.”

Volker Hoffmann, economist, Swiss Federal Institute of Technology Zurich

“What’s terribly unfair about climate change is that it affects those most who are the least to blame and whose island paradises and rich hunting grounds disappear into the sea, while we set up snow machines and complain about ski slopes being too hard-packed. Our lifestyle causes collateral damage, and we’d rather ignore that.”

Christian Pohl, environmental scientist, Swiss Federal Institute of Technology Zurich and Swiss Academies of Arts and Sciences, Bern

“If we want to get an idea of the climate awaiting us in the near future, we have to journey back in time to extremely warm episodes in Earth’s history.”

Helmut Weissert, geologist, Swiss Federal Institute of Technology Zurich

GLOSSARY

ADVECTION
Large-scale transport of air masses (both warm and cold air) by horizontal winds. Advection is the main cause of the typically unsettled weather conditions found in unstable climate zones such as the (temperate) west wind zone, as it frequently carries in air masses from other zones. Convection is the opposite process, referring to vertical air movements.

AEROSOL
Solid and liquid particles in the atmosphere such as dust, smoke, and salt crystals. Extremely small aerosol particles, measuring only 0.0001 to 10 micrometers, float in the air and can absorb and reflect radiation.

AGGREGATE STATE
The physical state of a substance: solid, liquid, or gas. When a substance (such as water) changes from one aggregate state to another, heat is either released (condensation, freezing) or consumed (melting, sublimation, evaporation).

AGROFUELS
More commonly referred to as biofuels. A distinction is made between three different generations. The first generation is fuel made from the oils, sugars, and starches of cultivated plants such as corn, soy, and sugar cane. Only parts of these plants are used for fuel production. These fuels are very controversial because their cultivation competes with the production of food. Moreover, the ecobalance of several types of agrofuel, for example those made from corn or rape, is unfavorable. In the production of second-generation agrofuel, most of the plant material is used, sometimes even the cellulose fibers that are difficult to separate out. These include wood and agricultural remnants as well as catering wastes. The third generation is made from algae. Algae have a significantly higher energy yield per unit area and do not compete with food production. However, the production of second and third generation fuel is technically much more demanding and cost-intensive.

ALBEDO
The fraction of incident light reflected by an object or the Earth's surface, in other words, a measure of reflective power. Whereas fresh fallen snow reflects up to 90 percent of incoming light, forests and fields reflect only 10 to 20 percent. Albedo is an important factor in the Earth's radiation balance.

ANNEX I COUNTRIES
In the UN Framework Convention on Climate Change, industrial nations are listed by name as Annex I countries, with the EU being listed as one entity only. When the term Annex I Countries is used, it is in reference to these listed industrialized countries. Countries in transition (so-called economies in transition) are also included in Annex I, i.e. all former Eastern Bloc countries that were in a state of transition from a planned economy to a free market economy in 1990.

ANNEX B COUNTRIES
In the Kyoto Protocol, Annex I countries (except Turkey and Belarus) are listed in Annex B. Each of these countries has a quantified emissions limit to comply with by the end of the first commitment period, lasting from 2008 to 2012.

ANNUAL MEAN TEMPERATURE
The average temperature at a given location over a period of one year.

ANTHROPOGENIC
Scientific term (adjective) for man-made, or caused by human activity.

ARIDITY
Meteorological term for the ratio of evaporation to precipitation. An area is deemed arid when, on a yearly average, evaporation is higher than the amount of rainfall or snowfall.

ATMOSPHERE
The protective envelope of gases surrounding the Earth. The atmosphere is divided into different layers and becomes increasingly thinner as altitude increases. The layer nearest to the Earth's surface is called the troposphere, which is between 5 kilometers in height at the poles and up to 17 kilometers at the Equator. The next layer is the stratosphere (10 to 50 kilometers), followed by the mesosphere (50 to 80 kilometers), and finally the thermosphere at an altitude of 80 to 700 kilometers.

ATMOSPHERIC WEST WIND DRIFT
→ Planetary frontal zone

BACK RADIATION
Thermal radiation directed back toward Earth from the atmosphere. Clouds and atmospheric water vapor absorb and radiate back most of Earth's outgoing long-wave thermal radiation.

BIOFUELS
→ Agrofuels

BIOMASS
The total mass of all living species in a given system.

BIOSPHERE
Term used to refer to all of the Earth's ecosystems combined; the areas inhabited by animals, plants, and other living organisms.

CARBON DIOXIDE (CO_2)
Carbon dioxide is an odorless, colorless gas. It is present in the Earth's atmosphere in small quantities. Carbon dioxide is produced through combustion and respiration and is consumed in photosynthesis to produce carbohydrates. The increase of carbon dioxide in the atmosphere since the beginning of industrialization is considered to be the principle reason for the global rise in surface temperatures caused by the greenhouse effect.

CARBON SINKS AND STORES

Geologists use the term sink for a reservoir that temporarily or permanently absorbs and stores a substance, for example carbon. Whereas a carbon store is static, a sink is dynamic, removing carbon from the carbon cycle. Oceans and young forests are important sinks as they absorb carbon from the atmosphere. Most soils are carbon stores, but not necessarily sinks. The same applies to older forests. Stores can turn into sources of carbon if they lose their storage capacity and start to release carbon, for example, when drained peat bogs start to decompose or when forests burn. → LULUCF

CFC

Abbreviation for chlorofluorocarbon. For a long time CFCs were, and sometimes still are, used as gas propellants in products such as hair spray, shaving creams, and insulating foams. They have a damaging effect on the ozone layer, leading to its depletion. Moreover, CFCs also contribute to the greenhouse effect by absorbing the Earth's thermal radiation. They absorb up to many thousand times more heat than CO_2 and are highly persistent in the atmosphere. For example, the half-life of CFC 115 in the atmosphere is approximately 1,700 years. After that time, only half of its molecules will have broken down.

CIRRUS CLOUDS

Fine, featherlike high ice clouds at altitudes of six to thirteen kilometers in blue skies. They are of key importance to the Earth's radiation balance.

CLEAN DEVELOPMENT MECHANISM (CDM)

A flexibility mechanism included in the Kyoto Protocol. It allows industrialized countries to fulfill their reduction obligations by making financial contributions to projects that help developing countries reduce their greenhouse gas emissions. A share of the proceeds of CDM activities feeds into the Adaption Fund. → Emissions trading

CLIMATE

Typical meteorological conditions observed and recorded over longer periods in a given place. The term weather, by contrast, describes short-term or daily atmospheric conditions.

CLIMATE CONFERENCES

For an overview of the most important climate conferences → p. 568.

CLIMATE ELEMENTS

Climate elements are meteorological conditions that can be measured. All climate elements combined determine the climate. Precipitation, temperature, evaporation, wind, atmospheric pressure, humidity, radiation, and cloud cover are important climate elements.

CLIMATE FACTORS

Climate factors refer to spatial features that can have an influence on the climate. Important climate factors include latitude, longitude, exposure to wind and sun, and population density. Human activity also influences the Earth's climate and is therefore a climate factor.

CLIMATE SYSTEM

The climate system is the joint action of climate elements in the atmosphere and their interactions with the hydrosphere, biosphere, and land areas. The atmosphere is the most obvious part of the climate system and the one that reacts most rapidly.

CLIMATE ZONES

Climate zones are extended regions on Earth that have a similar climate. Differences in solar radiation result in specific regional atmospheric circulation patterns, which is why the Earth has different climate zones.

CO_2 EQUIVALENT (CARBON DIOXIDE EQUIVALENT)

Various greenhouse gases differ in how much heat they trap in the atmosphere (global warming potential). To make its greenhouse impact comparable, the emissions of a given gas are multiplied by its global warming potential (GWP), a figure stating how much more heat a ton of the gas traps than one ton of CO_2. The result is in tons of CO_2 equivalent, i.e. its greenhouse effect is equal to that of the same tonnage of CO_2 emissions.

COMMITMENT PERIOD

The Kyoto Protocol provided for a first commitment period from 2008 to 2012. During this period, industrial nations were to reduce their greenhouse gas emissions by a defined percentage from 1990 levels. The 1997 Kyoto agreement also stipulated that considerations regarding obligations for the following period, beginning in 2013, were to be initiated by 2005 at the latest. By the end of 2010, however, signatories had not yet been able to agree on new reduction targets.

CONDENSATION

Condensation is when water vapor, a gas, becomes a liquid and turns into drops. The process of condensation always releases energy.

CONDENSATION HEAT

Energy released as heat during the transition of water vapor to its liquid aggregate state.

CONDENSATION NUCLEI

Cloud condensation nuclei (CCNs) are also called cloud seeds. They are tiny particles that float in the atmosphere (for instance dust or soot), facilitating the transition of vapor in the air to the liquid state by supplying a surface for droplets to condense on, form clouds, and fall as rain or snow.

CONTINENTAL ICE SHEETS

A huge expanse of permanent ice that almost completely covers a land area, such as Greenland or Antarctica. Only isolated nunataks (ridges, mountains, or peaks) protrude from the ice sheet. Also referred to as continental glaciers.

CONVECTION

Convection is the vertical motion of air masses of different densities. As air above the ground is warmed, it rises because it becomes less dense (lighter) than the surrounding air. Conversely, cold, dense air from higher altitudes sinks. Convection is an important process in the exchange of heat, mass, and momentum, and the formation of cumulus clouds.

CONVERGENCE

In meteorology, this term is used to refer to flow situations in which air parcels approach each other. For reasons of mass conservation, horizontal convergence is compensated by vertical divergence. As a consequence, convergence close to the Earth's surface leads to uplift and to the formation of clouds and precipitation. This is the case in low-pressure systems. Weather fronts are also convergence lines.

CONVERGENCE, INTERTROPICAL

→ Intertropical Convergence Zone

COP

Conference of Parties (COP) is a meeting of all states that are parties of the UN Framework Convention on Climate Change (UNFCCC). COPs take place once a year, usually between October and December. The first conference, COP1, was held in Berlin in the spring of 1995 shortly after the UNFCCC went into effect. All decisions must be reached unanimously. Since the Kyoto Protocol went into effect, the meetings of the parties to the Kyoto Protocol (Meeting of Parties or MOP) take place at the same time as the COPs. The first meeting, MOP1, was held together with COP11 in Montreal at the end of 2005. At the MOPs, the parties to the Kyoto Protocol make decisions concerning all aspects of the climate agreement. → p. 568 Climate Conferences

CORIOLIS EFFECT

The Coriolis effect is a force that acts on all bodies moving freely within a rotating reference frame (the spin of the Earth). It is a fictitious force because it requires bodies to be in motion. Due to the Coriolis effect, currents are deflected at a 90-degree angle to the direction of movement, to the right in the Northern Hemisphere and to the left in the Southern Hemisphere. As a result, low pressure systems rotate counterclockwise in the Northern Hemisphere and clockwise in the Southern Hemisphere.

CRYOSPHERE

The surface of the Earth covered by snow or ice, plus areas with frozen ground, including permafrost regions.

DESERT

A land area with sparse to nonexistent vegetation cover, characterized by a lack of water (dry deserts) or a lack of heat (cold deserts). Climate factors and topography both contribute to the occurrence of deserts. Geologists therefore differentiate between mid-latitude deserts, trade-wind deserts, coastal deserts, rain-shadow deserts, and continental deserts.

DIFFUSE SKY RADIATION

Term for all electromagnetic radiation from the sky that is not direct solar radiation. Diffuse sky radiation is solar radiation that has been scattered by clouds, dust, and aerosols, reaching the Earth's surface from all directions. Scattering is also responsible for the blue color of the sky.

DIRECT SOLAR RADIATION

The portion of solar radiation that penetrates the atmosphere and reaches the Earth's surface unhindered. It is the fraction of total insolation that is neither scattered (→ Diffuse Solar Radiation), reflected by clouds, nor absorbed, for instance, by the ozone layer.

ECCENTRICITY

In astronomy, eccentricity refers to a slight variation in the Earth's orbital path around the sun. It changes from a more circular to a more elliptical form within a 95,000-year cycle. Along with the Earth's → precession and changes in its axial tilt, these orbital variations influence the Earth's radiation balance.

EL NIÑO

Repeated anomaly in ocean surface temperatures in the tropical eastern Pacific and warm phase of the → ENSO. During an El Niño period, ocean surface temperatures off the coast of South America and in the eastern and central equatorial Pacific are significantly higher than usual. The consequences are heavy rainfall in the coastal desert regions of South America and diminished fish stocks off the South American west coast due to the lack of inflow of nutrient-rich deep waters. At the same time, Southeast Asia and eastern Australia are affected by droughts. El Niño phases are accompanied by high pressure conditions in the tropical western Pacific (→ Southern Oscillation). → La Niña is the name given to conditions characterized by cooler ocean surface temperatures.

EMERGING NATIONS

A group of states, not exactly defined, that have been classified as developing nations in the past, but that no longer fit that description due to their growing economic strength. These countries include Brazil, India, China, and South Africa. Together with Russia, the emerging nations form a political organization called BRICS. Within the framework of the United Nations, emerging nations would like to continue to be considered developing nations. Many of them are members of the Group of 77 and China. However, according to the new climate regime whose basic principles were laid out at the COP13 (13th Conference of the Parties to the UN Framework Convention on Climate Change) in Bali, they will be subject to meeting more obligations than poorer nations will.

EMISSIONS

Solid, liquid, and gaseous substances, sonic waves, or electromagnetic radiation released into the environment from a given source. The reception of pollutants, sound, or radiation by humans or the environment is called immission.

EMISSONS TRADING

This refers to the buying and selling of emissions permits or emission certificates. A certificate gives its holder the right to emit one ton of CO_2. The market functions according to the cap-and-trade system. This means that a state or a company is given permission to release a limited amount of emissions—this is the cap. These permits may be traded. Depending on the market price of a ton of CO_2, it may be less expensive for a company to buy emissions rights on the market than to reduce the amount of greenhouse gases it releases. For other companies it could make more financial sense to invest in measures to reduce its own greenhouse gas emissions. If a company reduces its emissions and has emission permits to spare, it can profit from selling the excess amount of emissions permitted in the form of certificates. Companies in the EU are obliged to participate in the European Union Emissions Trading Scheme, the largest trading system worldwide. There are also national CO_2 markets and exchanges. The largest trading platform is the European Climate Exchange, which opened in London in 2008. There are also CO_2 exchanges in Amsterdam, Oslo, Utrecht, and Paris.

ENSO

Acronym for the repeated climate pattern called El Niño Southern Oscillation. The term describes changes in surface water temperatures in the tropical eastern Pacific Ocean, called El Niño/La Niña, and variations in air pressure in the tropical western Pacific (→ Southern Oscillation). The warm El Niño phase is accompanied by high air pressure conditions and the cool La Niña phase is accompanied by low air pressure conditions in the western Pacific (or Southern Oscillation). Short-term climate swings and changes in circulation patterns related to ENSO can influence weather conditions throughout the entire world. → El Niño

EROSION

Term describing the breakdown and removal of soil and rock from one place on the Earth's surface to another.

GEF

The Global Environment Facility (GEF) was established by the World Bank in 1991 as a mechanism to finance environmental projects. It was restructured in 1994 and turned into an independent organization to increase developing nations' influence in decision-making processes. No institution today provides more funding for global environmental protection projects than the GEF. It serves as a financial mechanism for the UN Framework Convention on Climate Change and other environmental conventions.

GLACIAL

An adjective denoting the effect of a glacier. For example, "glacial erosion" means the breakdown and shifting of underlying ground layers such as bedrock by the ice of a glacier. A glacial (noun) or a glacial period is the scientific term for a cold period within an ice age. Glacials alternate with warm → interglacials.

GLOBAL RADIATION

The total amount of electromagnetic radiation that reaches the Earth's surface. Global radiation is the sum of direct and diffused solar energy.

GREENHOUSE GAS

Gases that contribute to the warming of the global climate by trapping long-wave radiation in the atmosphere. These gases include carbon dioxide, methane, nitrous oxide, ozone, and numerous gases containing sulfur and fluoride. The most important greenhouse gas is water vapor.

GROUP OF 77 AND CHINA, THE

The Group of 77 and China was established at the first United Nations Conference on Trade and Development (UNCTAD) in Geneva in 1964. At the time it comprised seventy-seven developing nations. It has since grown to encompass 131 member states, making it the largest intergovernmental organization of developing nations within the United Nations. The aim of the group is to enable the countries of the South to promote common economic interests and strengthen their negotiating capacity. → Emerging Nations

HADLEY CELLS

Year-round closed circulation systems to the north and south of the Equator, located between the equatorial band of low air pressure known as the → Intertropical Convergence Zone (ITCZ) and the subtropical high pressure belt. Warm air masses rise up in the ITCZ and then flow poleward north and south at a height of about ten to fifteen kilometers. When they reach the band of high air pressure in the subtropics, they sink to the surface, turning into warm, dry trade winds that flow back toward the Equator.

HURRICANE

A tropical cyclone or storm in the western Atlantic. Hurricanes usually develop in the Caribbean, the Gulf of Mexico, and the tropical North Atlantic at ocean surface temperatures above 27 degrees Celsius (80°F). They can cause severe damage as far as the Southwest of the United States. According to the Saffir-Simpson scale, a category 5 hurricane has sustained wind speeds that exceed 250 kilometers per hour. The Saffir-Simpson scale measures the wind speeds of hurricanes to rate their intensity and give an estimate of potential damage. Outside of the tropics, storms with wind speeds exceeding 117 kilometers per hour are also referred to as hurricanes; their wind speeds can even exceed 200 kilometers per hour. Hurricane Lothar, a low-pressure air system over Europe, attained wind speeds exceeding 270 kilometers per hour when it hit southern Germany, northern France, and Switzerland on December 26, 1999, causing major damage to forests.

HYDROSPHERE

The total of all water bodies on Earth, including oceans, inland seas, lakes, rivers, and groundwater, as well as ice and snow, and water vapor in the atmosphere.

INCLINATION

The tilt angle, often to a horizontal plane. For example, the angle of the Earth's magnetic field can be made visible by the angle that a freestanding magnetic needle makes with any horizontal surface. The angle, or inclination, is given in degrees.
In astronomy, inclination is the angle between one plane that is used as a reference and another plane or axis. It can be used to describe the shape and direction of a planet's orbit.

INDUSTRIALIZATION
Term for the period during which human society developed from an agrarian economy to an economic system characterized by the mechanized mass production of goods, a growing service industry, and the development and increase of motorized traffic. This kind of economy depends on the burning of fossil fuels such as coal, oil, and gas for the production of energy. Some climate scientists consider 1750 to mark the beginning of industrialization while others prefer to use 1850.

INFRARED RADIATION
Also referred to as thermal radiation. It is electromagnetic radiation with a wavelength longer than that of visible light (beyond 740 nanometers). Incoming solar radiation is absorbed by the Earth's surface and reflected as infrared radiation.

INTERGLACIAL
A warm interval that separates two → glacials. Due to their much warmer climate, they cause the ice shields and glaciers on Earth to shrink. The last interglacial before the present one, the Eemian interglacial (Riss-Würm interglacial in the Alps, Sangamonian Stage in North America), occurred between the Vistulian glacial (Würm glacial in the Alps, Wisconsin glacial in North America) and the earlier Saalian glacial (Riss glacial in the Alps, Illinoian glacial in North America). Even earlier interglacials were the Holstein interglacial (Mindel-Riss in the Alps, Pre-Illinoian in North America) and the Günz-Mindel interglacial. Different names are used for the same interglacials in Great Britain and South America. → p. 567 Table Timeline of Geologic Eras

INTERTROPICAL CONVERGENCE ZONE (ITCZ)
The Intertropical Convergence Zone is a low-pressure trough near the Equator that encircles the entire globe. This is where the northeasterly and southeasterly trade winds converge. Low air pressure in the Intertropical Convergence Zone is a result of air masses rising due to intense solar radiation. The warm air rising here carries a lot of moisture, leading to the formation of towering thunderstorm clouds and near-daily thunderstorms. The location of the ITCZ does not necessarily coincide with the Equator. Over land areas, it moves north and south with the sun's zenith point. The ITCZ frequently separates into two ITCZs, one in the north and one in the south.

IPCC
Intergovernmental Panel on Climate Change. A scientific advisory body created by the UN Environment Programme (UNEP) and the World Meteorological Organization (WMO). Its task is to review and assess the most recent scientific, technical, and socioeconomic information relevant for understanding the risk of human-induced climate change.

ISOTOPES
Isotopes are variants of a chemical element that have different numbers of neutrons in the nucleus. The radioactive decay of 14C, a natural radioactive form of carbon, is used for → radiocarbon dating.

KYOTO PROTOCOL
The Kyoto Protocol was adopted in Kyoto on December 11, 1997 as an additional protocol to the United Nations Framework Convention on Climate Change; it has been ratified by nearly all member parties. The protocol went into force on February 16, 2005, and its first commitment period expires at the end of 2012. The Kyoto Protocol marks the first time that signatories from industrialized nations have agreed to binding targets for their emissions of greenhouse gases. As of now, 192 states have signed the protocol. The United States is not party to the Kyoto Protocol.

LA NIÑA
Cool phase of the → ENSO and counterpart to El Niño. During a La Niña period, surface water temperatures of the Pacific off the coast of South America and along the Equator are significantly lower than usual. The result is that the coastal regions of South America are drier than usual. At the same time, Southeast Asia and eastern Australia experience above average rainfall.

LIGHT
Light is the part of electromagnetic radiation visible to humans. White sunlight is made up of the light of different wavelengths that the human eye perceives as color (the spectral colors of the rainbow). Visible light ranges in wavelength from 400 to 760 nanometers. In the continuous electromagnetic spectrum, wavelengths shorter than visible light are called ultraviolet radiation, while wavelengths longer than visible light are called infrared radiation.

LONG-WAVE RADIATION
In climatology, this term refers to wavelengths that are longer than 400 nanometers.

LULUCF
Acronym for land use, land-use change, and forestry. LULUCF has played an important role in climate policy since the Kyoto Protocol was adopted. The UN Framework Convention on Climate Change had already committed member states to conserve and enhance sinks and reservoirs of greenhouse gases, including biomass, forests, and oceans, as well as other terrestrial, coastal, and marine ecosystems. But the Kyoto Protocol explicitly lists LULUCF, land use for agricultural purposes, land-use change, and forestry, as activities that must be taken into consideration to a certain extent by industrial nations in the fulfillment of their obligations to reduce greenhouse gases.

MACROCLIMATE
Macroclimate refers to the climate of a very large area. The characteristics of a macroclimate depend mainly on global atmospheric circulation.

METHANE (CH4)
Methane is a combustible and gaseous hydrocarbon. It is a major greenhouse gas.

MONSOON

A wind system in South and Southeast Asia characterized by a shift in the prevailing wind direction every six months. Monsoons are caused by both seasonal changes in continental pressure patterns over Asia and the seasonal shifts of global wind systems. The southwest summer monsoon brings extremely heavy rainfall to the countries concerned.

MONTREAL PROTOCOL

Multilateral United Nations Protocol on Substances that Deplete the Ozone Layer. It is binding under international law and went into force on January 1, 1989. By ratifying the treaty, member states agreed to control the emission of certain substances that lead to the depletion of the ozone layer, with the ultimate goal of phasing out these substances. The provisions of the protocol were later reviewed and tightened several times. The agreement, which has been ratified by 195 countries, is considered an example of successful international environmental policy.

NITROGEN OXIDES

General term for nitric oxide (NO) and nitrogen dioxide (NO_2). The colorless NO is formed predominantly in combustion processes. It immediately oxidizes when it comes into contact with air and produces NO_2, a brownish gas that has a biting odor. Nitrogen oxides and hydrocarbons react in the presence of sunlight to form ozone, a major contributor to summer smog conditions.

NITROUS OXIDE (N_2O)

Nitrous oxide is a chemical compound also known as laughing gas. It is an important greenhouse gas that is produced by combustion or by soil bacteria, mainly where fertilizers are used excessively in agriculture.

OBLIQUITY

In the solar system, the Earth's orbital path around the sun is called the ecliptic plane. The Earth's rotation axis is not perpendicular to the ecliptic plane, but tilted at an angle of about 23.5 degrees (23°27'). The axial tilt is also called the obliquity of the ecliptic.

OECD

The OECD (Organization for Economic Co-operation and Development) is an organization of thirty-four countries, most of which have high per capita income. They are considered to be developed countries.

OPEC

Acronym for Organization of Petroleum Exporting Countries. These countries account for about 40 percent of worldwide oil production and hold more than three-fourths of the world's oil reserves. Of the world's ten biggest oil producing countries, only Saudi Arabia, Iran, Kuwait, Venezuela, and the United Arab Emirates are members of OPEC. The goal of the organization is to create a monopolized oil market, and to protect its members against fluctuations in the world market price of oil by setting quotas on crude oil extraction and production.

OZONE (O_3)

Ozone is a gas that has a specific sharp odor. Its molecules are made up of three oxygen atoms (O_3). Ozone is produced in the atmosphere when ultraviolet solar radiation reacts with molecular oxygen (O_2). Ozone concentrations are highest at altitudes of 20 to 30 kilometers. This area is referred to as the ozone layer. This chemically reactive gas is also produced in the lower atmosphere close to the Earth's surface when nitrogen oxides and hydrocarbons react under intense solar radiation. High concentrations of ozone in ambient air are the cause of respiratory problems.

OZONE LAYER

A layer in the stratosphere at an altitude of 20 to 30 kilometers with high levels of ozone. The ozone layer absorbs harmful solar UV radiation, making life on Earth possible.

PALEOCLIMATOLOGY

The study of the history of Earth's climate. Paleoclimatology reconstructs climate developments and events in the past by using climate archives such as ice cores, sediments, and tree rings.

PEAK OIL

Peak oil is the point at which the maximum rate of oil extraction has been reached worldwide. Once the maximum has been reached, the global rate of oil extraction goes into decline. The exact time of peak oil is difficult to determine, however, as it depends on both physical conditions and the demand for oil. Some scientists believe peak oil has already occurred or is imminent; others believe it is twenty years away.

PEDOSPHERE

The pedosphere is the top layer of the Earth's surface, forming a boundary between the atmosphere and underlying bedrock. The pedosphere is where soils are formed.

PERMAFROST

Permafrost is predominant in polar and alpine regions (permafrost regions), where temperatures are so low that the soil remains permanently frozen. Thawing in summer occurs only in a thin upper layer, which ranges from several decimeters to a few meters in thickness. As the underlying strata remain frozen, water can not infiltrate the ground, resulting in widespread waterlogging.

PLANETARY FRONTAL ZONE

The latitude belt that comprises the main air mass boundaries, such as the → polar front. The planetary frontal zone is located between the warm high pressure subtropical belt and the cold subpolar low-pressure belt, lying between 35 and 60 degrees latitude in both the Northern and the Southern Hemispheres. Under the influence of the Coriolis effect, westerly winds prevail in this zone. At higher altitudes, west winds are very strong (jet streams). Because of the constant confrontation of cold polar air and warm tropical air in the planetary frontal zone, it is meteorologically very dynamic, with frontal systems continually forming and dissipating.

PLANKTON
Small plant or animal organisms that drift in water and frequently are not able to move independently or only to a limited extent. Plankton is primarily carried by water currents and is an important part of aquatic food chains.

PLATE TECTONICS
A theory based on the concept of continental drift, explaining the structure of the Earth's crust and the formation and shifting of continents and oceans over millennia. The theory assumes that the solid surface of the Earth is made up of large, mostly rigid plates that gradually and passively move as a consequence of flow processes in the Earth's mantle. Some accompanying processes are sea-floor spreading, mountain building, and volcanic and earthquake activity.

POLAR FRONT
Boundary between tropical and polar air masses, located at 40 to 50 degrees latitude in the north and south in winter, and 60 to 70 degrees latitude in the north and south in summer. This is where the low pressure systems develop which govern weather and climate conditions in temperate zones.

POLAR ZONE
General term for the climate zones in the polar regions north and south of the temperate zones. The polar zones extend south from the Antarctic Circle in the Southern Hemisphere and north from the Arctic Circle in the Northern Hemisphere. The ice deserts of the Arctic and the northern parts of Scandinavia, Siberia, Canada, and Alaska are in the polar zone of the Northern Hemisphere. In the Southern Hemisphere, the only land area in the polar zone is Antarctica.

PRECESSION
Precession in astronomy is a term that describes the spinning, top-like (gyroscopic) wobble of the Earth's axis. A complete revolution of this spinning movement takes 19,000 to 23,000 years. Precession only influences the position of the Earth's axis, but not its tilt angle. Global climate is influenced by precession, → eccentricity, and changes in the axial tilt combined.

RADIOCARBON DATING
Also known as carbon dating or 14C dating, this is a method used to determine the age of materials that contain carbon, such as wood, bones, limestone fossils, and so forth. The method is based on the amount of carbon-14, a radioactive isotope, detected. In the atmosphere the ratio of 14C to 12C is always the same because 14C is continually formed by cosmic radiation from nitrogen isotopes in the air. This ratio is also reflected in substances and living organisms into which carbon is incorporated at exactly this ratio. If further absorption of carbon from the atmosphere is prevented, for instance when animals or plants die, the ratio of 14C to 12C diminishes steadily as 14C decays (radioactive decay). The half-life of 14C is 5,730 years, i.e. half of the amount of 14C has disappeared after that time. By measuring the ratio of the two carbon isotopes, the time since the absorption of atmospheric carbon came to an end can be determined, indicating the age of the material. The age of finds as old as 50,000 years can be dated in this way.

RAINFOREST
1. Term referring to the forest ecosystem found in moist tropical regions. These forests experience year-round precipitation with more than 2,000 millimeters of rainfall per year and average annual temperatures exceeding 18 degrees Celsius (64.4°F). Tropical rainforests are home to an enormous variety of animal and plant species. They are typically divided into layers. The top layer is more than 50 meters above the forest floor and consists of very tall trees. These rainforests occupy roughly one-third of the world's forest areas.
2. Rainforest is also the term applied to forests in temperate regions that have developed lush vegetation under conditions of extremely high levels of moisture (e.g. fog) and precipitation.

REDD/REDD+
The acronym REDD stands for "Reducing Emissions from Deforestation and Forest Degradation." As the protection of tropical forests within the framework of the Kyoto Protocol is only possible by means of → Clean Development Mechanism emission-reduction projects, the UN has been exploring new options since 2005. The resulting proposal, REDD+, which was presented at COP13 (13th Conference of the Parties to the UN Framework Convention on Climate Change) in Bali, is an important element of the Bali Roadmap. In addition to the aims set out in REDD, it includes the promotion of conservation and the sustainable management of forests, and the systematic enhancement of carbon stores.

REFLECTION
Reflection is the process by which a surface reflects incoming electromagnetic waves such as light rays. There are two different kinds of reflection: diffuse reflection which refers to the reflection of rays at many angles from a rough or irregular surface; and specular, or mirror-like reflection, which is the reflection of rays from smooth surfaces such as water and snow.

SEASONAL WEATHER
Recurring sequence of meteorological phenomena within an annual cycle. Seasonal weather describes the typical progression of weather over a period of several days or weeks in an extended area. Seasonal weather is greatly influenced by large-scale meteorological conditions, mainly the typical distribution of high and low air pressure systems.

SEMI-ARID CLIMATE
A semi-arid climate describes a region with generally low precipitation and vegetation dominated by grasses and shrubs. During three to five months of the year, however, the amount of precipitation received is higher than the amount of water that evaporates.

SHORT-WAVE RADIATION
In climatology, this term refers to radiation with wavelengths below 400 nanometers.

SINKS AND STORES
→ Carbon sinks and stores

SOLAR CONSTANT

The amount of radiation energy received from the sun per unit of time per unit of area at the outermost layer of the atmosphere. It is the power received by a theoretical surface perpendicular to the incoming rays and at Earth's mean distance from the sun. The average value is about 1,367 watts per square meter. It is assumed to be relatively constant over hundreds of years, but fluctuates by 7 percent within the course of a year due to the Earth's varying distance from the sun.

SOUTHERN OSCILLATION

Periodic fluctuations in atmospheric pressure distribution over the southwest Pacific, a component of the El Niño Southern Oscillation (→ ENSO). The Southern Oscillation Index (SOI) measures the Southern Oscillation's air pressure distribution by computing variations in the surface air pressure difference between Darwin, Australia and Tahiti. If the SOI is negative, as is the case during an El Niño phase, air pressure in Tahiti is lower than in Darwin. The opposite, a positive SOI value, indicates a La Niña phase. The Southern Oscillation is connected to the → trade winds and the → Walker Circulation.

STERN REPORT

The Stern Review on the Economics of Climate Change is the full title of this report, published in October of 2006. It deals with the economic consequences of global warming. The British government commissioned Nicholas Stern, who formerly held the position of World Bank chief economist, to compile the report. Its main conclusion is that the prevention of global warming makes more economic sense than dealing with its consequences.

STRATOSPHERE

A layer in the Earth's atmosphere. It is located above the troposphere, the layer next to the surface in which weather phenomena develop and take place. The lower limit of the stratosphere lies at an altitude of about 17 kilometers at the Equator and 8 kilometers at the poles. The stratopause, the boundary between the stratosphere and the layer above, the mesosphere, is situated at an average altitude of 50 kilometers above the Earth's surface. Temperatures at this elevation are about 0 degrees Celsius (32°F). In the lower part of the stratosphere, the temperature is a relatively constant minus 55 degrees Celsius (-67°F). The ozone concentration is highest at altitudes of 20 to 30 kilometers—this area is referred to as the ozone layer. The stratosphere is the layer in which ozone absorbs short UV waves from space.

SUBTROPICAL RIDGE, THE

Or subtropical high pressure belt, a belt of constant high atmospheric pressure between the 20th parallel and 40th parallel in both the Northern and Southern Hemispheres. These highs are part of the trade winds and → Hadley cell circulation pattern. Their locations vary seasonally depending on the position of the sun, moving poleward during summer and equatorward in winter. The subtropical ridge of the Northern Hemisphere is made up of high pressure systems such as the Azores High and the Bermuda High, important weather factors respectively in Europe and on the eastern coast of North America.

SUNSPOT ACTIVITY

Sunspots are temporary dark spots on the photosphere of the sun that are 1,000–2,000 Kelvin cooler than the surrounding area. They are caused by intense magnetic activity. The number of sunspots varies within an irregular cycle of eleven years.

TEMPERATURE

A physical property of matter, indicating hot and cold. Temperatures are commonly given in degrees Celsius, or in degrees Fahrenheit in the United States. The Kelvin (K) temperature scale is used to indicate temperature differences and absolute temperatures in physics, chemistry, and astronomy. The conversion formula is degrees Celsius = degrees Kelvin −273.15. Absolute zero is defined as 0 Kelvin.

TERRESTRIAL

Pertaining to, or originating from the Earth or a land area (Latin: terra = earth).

TIPPING POINTS

Tipping points are especially critical points in complex systems. When a tipping point is reached, it can trigger unexpected and far-reaching changes with unpredictable and often irreversible consequences.

TORNADO

Also called a twister or a cyclone. This wind vortex usually affects only smaller areas, but it can be highly destructive. A tornado can develop over land areas during intense thunderstorms. A typical tornado is visible as a rotating funnel or tube which reaches from the storm cloud down to the ground. It ranges from 10 to a couple of 100 meters in diameter, with wind speeds in its core of up to 400 kilometers per hour.

TRACE GASES

Gases whose concentrations make up a minimal percentage of the total volume of the atmosphere. Trace gases are gases such as → carbon dioxide, argon, helium, krypton, xenon, ozone, → methane, hydrogen, sulfur dioxide, and → nitrogen oxides. Some of them are very important for the climate because they act as greenhouse gases.

TRADE WINDS

Steady surface winds that blow year round. Trade winds, also referred to as the trades, flow toward the Equator in both the Northern and Southern Hemispheres. They flow from the subtropical high-pressure belts in the direction of the equatorial low-pressure trough and are an important part of the tropical circulation system (→ Hadley cells). On their way toward the Equator, the → Coriolis effect and proximity to the ground cause the trades to blow from the northeast in the Northern Hemisphere, and from the southeast in the Southern Hemisphere. Trade winds are dry winds bringing little precipitation.

TROPOSPHERE

Lowest layer of the Earth's atmosphere where most weather phenomena occur. The troposphere extends to an approximate altitude of 17 kilometers at the Equator and up to 8 kilometers at the poles. Temperatures in the troposphere drop by about 0.6 degrees Celsius (1.1°F) for every 100 meters altitude. The next layer in the Earth's atmosphere is the troposphere; the boundary between the troposphere and the stratosphere is called the tropopause. The temperature of the tropopause above the Equator is minus 80 degrees Celsius (-112°F); above the poles it is minus 50 degrees Celsius (-58°F). Air pressure decreases from average sea-level values of 1,013 hectopascals to tropopause levels of 100 hectopascals over the Equator, and up to 400 hectopascals over the poles.

TYPHOON

Typhoon refers to tropical cyclones that develop in the northwestern Pacific and can cause severe devastation in East Asia, for instance in Japan and China.

UNITED NATIONS (UN)

The United Nations was founded by fifty-one nations in 1945 after World War II. The aims of the organization are to maintain world peace and security, and to improve living conditions throughout the world. Today every internationally recognized state is a member of the organization, making it truly global. Its most important body is the UN General Assembly. Other bodies include the Secretariat headed by the UN Secretary-General, the Security Council, the Economic and Social Council, and the International Court of Justice in The Hague. It has several additional institutions, for example the UN Environment Programme (UNEP), and specialized agencies such as the World Meteorological Organization (WMO) and the World Bank.

UNITED NATIONS FRAMEWORK CONVENTION ON CLIMATE CHANGE

The UN Framework Convention on Climate Change (UNFCCC or FCCC) is a treaty that was ratified at the Earth Summit in Rio de Janeiro in 1992. It went into effect in March 1994. The aim of the treaty is "to stabilize greenhouse gas concentrations in the atmosphere at a level that would prevent dangerous anthropogenic interference with the climate system," and for this to take place quickly enough "to allow ecosystems to adapt naturally to climate change, to ensure that food production is not threatened and to enable economic development to proceed in a sustainable manner." Current parties to the convention include 194 states and one regional economic organization. The treaty itself does not set any specific quantitative targets for the reduction of greenhouse gas emissions. It does however lay the legal basis for additional instruments such as the Kyoto Protocol.

VEGETATION ZONES

Vegetation zones are extensive land areas of similar vegetation. As plant life is governed largely by sun intensity, rainfall, and average temperatures, vegetation zones closely follow climate zones. Boundaries between different vegetation zones often deviate markedly from those of climate zones, however, because other factors such as topography, soil type, and water availability also influence plant life.

WALKER CIRCULATION

The Walker Circulation occurs along the Equator over the tropical Pacific Ocean. It is driven by the Pacific's varying surface temperatures and is linked to the → ENSO. Easterly trade winds near the Equator are part of this circulation, picking up moisture and bringing it to the western tropical Pacific. Here moist air is heated, rises, and sheds its moisture as rain. The now drier air flows back to the eastern tropical Pacific (toward South America) in westerly winds at higher altitudes, closing the cycle of the Walker Circulation. If fluctuations in atmospheric pressure of the → Southern Oscillation trigger an → El Niño phenomenon, the direction of the Walker Circulation is reversed.

WEATHER

Description of current atmospheric conditions and short-term meteorological developments. Weather is made up of elements such as clouds, precipitation, wind, temperature, and air pressure, and occurs exclusively within the troposphere. Seasonal weather, by contrast, is the typical recurrence of specific conditions within an annual cycle. The term climate refers to average meteorological conditions observed and recorded over periods of at least ten years.

WORLD BANK

The World Bank, along with the International Monetary Fund (IMF), was created in July 1944 by the founding members of the United Nations at the Bretton Woods Conference in the United States. At that time, its main priorities were monetary stability, reconstruction, and the rebuilding of economies in the wake of World War II. It is a special institution of the United Nations. Since the 1960s, its primary goal has been to reduce poverty and improve living conditions in poor countries. Industrial nations dominate the Board of Governors— although all 187 member states are represented, the weight of a country's vote depends on the size of its financial contribution to the bank. Two institutions of particular importance belong to the World Bank Group: the International Bank for Reconstruction and Development (IBRD) and the International Development Association (IDA).

TIMELINE OF GEOLOGIC ERAS

Era	Period		Epoch		Beginning (millions of years ago)
Cenozoic	Quaternary		Anthropocene*		0.00015
			Holocene		0.01
			Pleistocene		1.8
	Tertiary	Upper Tertiary (Neogene)	Pliocene	Upper Pliocene	3.6
				Lower Pliocene	5.3
			Miocene	Upper Miocene	11.6
				Middle Miocene	16.0
				Lower Miocene	23.3
		Lower Tertiary (Paleogene)	Oligocene		33.9
			Eocene		55.8
			Paleocene		65.5
Mesozoic	Cretaceous		Upper Cretaceous		99.6
			Lower Cretaceous		145.5
	Jurassic		Upper Jurassic		161.2
			Middle Jurassic		175.6
			Lower Jurassic		199.6
	Triassic		Upper Triassic		228.0
			Middle Triassic		245.0
			Lower Triassic		251.0
Paleozoic	Permian		Upper Permian		256.0
			Lower Permian		299.0
	Carboniferous		Upper Carboniferous		326.4
			Lower Carboniferous		359.2
	Devonian		Upper Devonian		385.3
			Middle Devonian		397.5
			Lower Devonian		416.0
	Silurian		Upper Silurian		423.0
			Lower Silurian		443.7
	Ordovician		Upper Ordovician		460.9
			Middle Ordovician		471.8
			Lower Ordovician		488.3
	Cambrian		Upper Cambrian		501.0
			Middle Cambrian		513.0
			Lower Cambrian		542.0
Precambrian	Proterozoic				2,500
	Archean				3,800

*In 2002, Nobel laureate Paul Crutzen proposed using the term Anthropocene for the new era in Earth's history characterized by human activity.

THE MOST IMPORTANT CLIMATE CONFERENCES

Year	Name	Place	Significance	Page
1979	World Climate Conference (WCC1), organized by the WMO	Geneva	The influence of climate on humankind is at the heart of talks. Governments are called upon to prevent anthropogenic changes to the climate that may harm the welfare of humankind. The conference notes that the cooling down of the climate in recent decades is comparable to natural episodes of cooling in the past, and that it also appears plausible that a further increase of CO_2 in the atmosphere, which seems likely to happen, will lead to a gradual warming of the lower atmosphere. The World Meteorological Organization (WMO) initiates the World Climate Programme (WCP).	456
1985	International Conference on the Assessment of the Role of Carbon Dioxide and of Other Greenhouse Gases in Climate Variations and Associated Impacts	Villach	This conference is held jointly by the WMO, the United Nations Environment Programme (UNEP), and the International Council for Science (ICSU). There is now little doubt that anthropogenic greenhouse gas emissions will cause global warming. Participants urge governments to take steps toward initiating a global climate convention. Indirectly, the conference triggers the founding of the Intergovernmental Panel on Climate Change (IPCC).	452 456 459
1988	Toronto Conference on the Changing Atmosphere: Implications for Global Security	Toronto	Scientists propose that greenhouse gas emissions be reduced by 20 percent, compared to 1988 values, by 2005.	
1990	2nd World Climate Conference (WCC2)	Geneva	The United Nations is called upon to draft an international climate convention. The first IPCC report is released shortly before the conference convenes. It notes that clear evidence of anthropogenic warming may not be available for ten or more years, given uncertainties and extreme variability in data.	457 460
1992	United Nations Conference on Environment and Development (UNCED), Earth Summit	Rio de Janeiro	The United Nations Framework Convention on Climate Change (UNFCCC), worked out during the previous two years, is ready for signing at UNCED.	457 464 *482* 506
1995	COP1 1st UN Climate Change Conference	Berlin	After the Convention comes into force in March 1994, the first Conference of Parties (COP) meets in Berlin. The mandate is to set up the guidelines for the first climate protocol, to be adopted within the next two years.	467 469 *483*
1997	COP3 3rd UN Climate Change Conference	Kyoto	The Kyoto Protocol is adopted. Many questions remain unanswered in detail and negotiations in following years are expected to find solutions. The conference bases its decisions on the second IPCC report (SAR) released late in 1995, which says that the balance of evidence suggests a discernible human influence on global climate.	468 470 474 476 *484*
2000	COP6 6th UN Climate Change Conference	The Hague	The conference breaks up without agreement and is continued six months later as COP6-bis in Bonn. Still up for discussion is the specific interpretation of the Kyoto Protocol.	*486* *487*
2001	COP6-bis Part 2 of 6th UN Climate Change Conference	Bonn	American President George W. Bush declares in the spring that the United States will not ratify the Kyoto Protocol. The conference is able to resolve many open issues and the definitive finalization of the Kyoto Protocol is within reach.	*488*

2001	COP7 7th UN Climate Change Conference	Marrakesh	The Marrakesh Accords enable the ratification of the Kyoto Protocol. The third IPCC report (TAR) is released in 2001. It estimates that there is at least a 66 percent probability that most of the warming of the last century is due to anthropogenic influence.	*489* 522
2005	COP11 11th UN Climate Change Conference	Montreal	The conference is held barely a year after the Kyoto Protocol enters into force. Simultaneously, Kyoto Protocol signatories convene for the first time at the first Meeting of Parties (MOP1), and negotiators begin thinking about the time after the protocol's first commitment period, which expires at the end of 2012.	474
2007	COP13 13th UN Climate Change Conference	Bali	The Bali Action Plan sets out guidelines for a future global climate regime that would prescribe emission reduction targets for the United States and major emerging countries. The conference follows the release of the fourth IPCC report (AR4), which notes that, with a probability of more than 90 percent, most of the global warming observed in the past fifty years can be attributed to human activity.	*490* 500 503
2009	COP15 15th UN Climate Change Conference	Copenhagen	The expectation that a new climate regime, as envisaged at the Bali conference, will be ready for signing in Copenhagen is not met. Frustration is high among its more than 40,000 participants.	*492* 503
2010	COP16 16th UN Climate Change Conference	Cancún	The main points of the Copenhagen Accord, formulated by major economies at the Copenhagen conference, are largely incorporated into the UN process, and the target to stay below a global warming of 2 degrees Celsius (3.6°F) is formally adopted.	*493* 510 526
2011	COP17 17th UN Climate Change Conference	Durban	In the Durban package, a second commitment period of the Kyoto Protocol was finally approved, together with a commitment for a legally binding global treaty to be adopted by 2015 and in force by 2020.	480 525

INDEX

FURTHER READING

Climate history, the climate system, and the consequences of climate change

David Archer, *Global Warming: Understanding the Forecast.* Hoboken, N.J., 2012.

Brian Fagan, *The Great Warming: Climate Change and the Rise and Fall of Civilizations.* New York, 2008.

Robert Henson, *The Rough Guide to Climate Change.* London, 2011.

Fred Pearce, *With Speed and Violence: Why Scientists Fear Tipping Points in Climate Change.* Boston, 2007.

Climate change—mitigation and adaptation

International Energy Agency (IEA), *Energy Outlook.* Washington, D.C., 2010.

International Energy Agency (IEA), *Global Gaps in Clean Energy.* Paris, 2010.

IPCC, *Synthesis Report (AR4).* Geneva, 2007. More detailed information can be found in the Summaries for Policymakers of the three AR4 Working Group Reports.

PriceWaterhouseCoopers, *100% Renewable Electricity: Roadmap to 2050 for Europe and North Africa.* London, 2010.

Mycle Schneider, *The World Nuclear Industry Status Report.* Paris and Berlin, 2009.

Henning Steinfeld et al., *Livestock in a Changing Landscape.* Washington, D.C., 2010.

Greenpeace International and European Renewable Energy Council (EREC), *Energy (R)evolution.* Amsterdam, 2010.

UNEP, *The Environmental Food Crisis.* Nairobi, 2009.

UNEP, EPO, and ICTSD, *Patents and Clean Energy, Final Report.* Munich, 2010.

Global climate politics

Shardul Agrawala, "Context and Early Origins of the Intergovernmental Panel on Climate Change." *Climatic Change* 39, no. 4 (1998), pp. 605–20.

Joseph Alcamo et al., *The Emissions Gap Report: Are the Copenhagen Accord Pledges Sufficient to Limit Global Warming to 2°C or 1.5°C? A Preliminary Assessment.* UNEP, Nairobi, 2010.

Daniel M. Bodansky, "The Emerging Climate Change Regime." *Annual Review of Energy and Environment* 20 (1995), pp. 425–61.

Andrew E. Dessler, *Introduction to Modern Climate Change.* Cambridge, UK, 2012.

Robbert H. Dijkgraaf and Lu Yongxiang, *Climate Change Assessments: Review of the Processes and Procedures of the IPCC.* InterAcademy Council, Alkmaar, 2010.

Andreas Fischlin et al. "Bestandesaufnahme zum Thema Senken in der Schweiz" [An inventory of sinks in Switzerland]. *Systems Ecology Report,* no. 29. Zurich, 2003.

Alan D. Hecht and Dennis Tirpak. "Framework Agreement on Climate Change: A Scientific and Policy History." *Climatic Change* 29, no. 4 (1995), pp. 371–402.

International Energy Agency (IEA), *CO_2 Emissions from Fuel Combustion: Highlights.* Washington, D.C., 2010.

Irving M. Mintzer and Leonard, J. Amber, eds., *Negotiating Climate Change: The Inside Story of the Rio Convention.* New York, 1994.

Nicholas Stern, *The Economics of Climate Change: The Stern-Review.* London, 2006.

Robert T. Watson et al., *Land Use, Land-Use Change, and Forestry: A Special Report of the IPCC.* New York, 2000.

Spencer Weart, *The Discovery of Global Warming.* Cambridge, Mass. and London, 2008.

Meles Zenawi and Jens Stoltenberg, *Report of the Secretary-General's High-Level Advisory Group on Climate Change Financing.* New York, 2010.

RELATED LINKS

Sustainability worldwide, Millennium Goals
www.worldwatch.org
www.weed-online.org

Environmental economics, 2,000-watt society, renewable energies
www.ecoeco.org
www.sternreview.org/uk
www.novatlantis.ch
www.energie-cities.eu
www.erneuerbare-energien.de
www.eurosolar.de
www.ren21.net

Major sources of data, latest findings in climate research
www.klimablog.ethz.ch
www.bafu.admin.ch/klima
www.umweltbundesamt.de/klimaschutz/index.htm
www.hamburger-bildungsserver.de (keyword Klima)
www.unep.org/ourplanet (online newspaper)

Major research institutes
Swiss Federal Institute of Technology Zurich (ETHZ), Department of Environmental Sciences
www.env.ethz.ch
Oeschger Centre Graduate School of Climate Sciences, University of Berne
www.oeschger.unibe.ch/index_de.html
Potsdam Institute for Climate Impact Research
www.pik-potsdam.de
Alfred Wegener Institute
www.awi.de
German Research Center for Environmental Health
www.helmholtz-muenchen.de
Wuppertal Institute for Climate, Environment and Energy
www.wupperinst.org

Other institutions
www.ipcc.ch
www.iea.org
www.energywatchgroup.org
www.energiekrise.de
www.worldcoal.org
www.globalsubsidies.org
www.waterfootprint.org
www.fao.org
www.redd-net.org/
reddpluspartnership.org
unfccc.int/2860.php

Finances and Fast Start climate financing
www.climatefundsupdate.org/
www.climatefundsupdate.org/fast-start-finance

LARS MÜLLER PUBLISHERS

WHO OWNS THE WATER?
Lars Müller, Klaus Lanz, Christian Rentsch, and René Schwarzenbach (Eds.)
ISBN 978-3-03778-015-2

"It is an aesthetic book, filled with pictures, and a fascinating read at the same time.... The aptly chosen texts and evocative illustrations convey factual and visual information in a refreshing way that is also amazing and frightening."
Quell Verlag für nachhaltiges Leben

THE FACE OF HUMAN RIGHTS
Walter Kälin, Lars Müller, and Judith Wyttenbach (Eds.)
ISBN 978-3-03778-114-2

"The editors have done justice to the issue in an outstanding manner. Their encyclopedic work illustrates the full story of human rights more vividly than this has ever been done before."
Der Bund

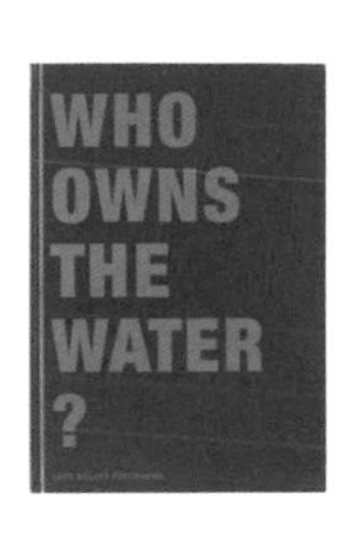

www.lars-muller-publishers.com

ABOUT THE AUTHORS

Heidi Blattmann
Heidi Blattmann holds a Master of Science ETH from the Swiss Federal Institute of Technology Zurich. Until the 1980s, she worked as a science journalist for the *Tages-Anzeiger* in Zurich and for Swiss Radio DRS. She also earned a Master of Arts in Social Sciences from the University of Zurich, with a focus on the general history of religion and cultural anthropology. From 1986 she worked as an editor at the *Neue Zürcher Zeitung,* first in the national section, later in the international section, and from 2002 until the end of 2008, she headed the newspaper's science section. She was appointed Honorary Councilor to the ETHZ in 2009. She has tracked the climate debate as a journalist since 1990; today she works as a freelancer.

Martin Läubli
Martin Läubli holds a degree in geography, environmental studies, and journalism from the University of Zurich. He has worked for eleven years at the *Tages-Anzeiger* in Zurich, focusing on environmental and energy issues. The first UN climate conference he experienced was in The Hague in 2000. Since then he has been the newspaper's correspondent on international climate negotiations. He has also become involved in teaching children about environmental issues, which has given him the opportunity to continue learning about cultural education and museum education. He recently developed a children's outdoor climate game in Oberengadin. He lives in St. Gallen.

Klaus Lanz
Klaus Lanz was born in Frankfurt am Main and holds a PhD in chemistry. As an environmental scientist and writer, he has been involved for more than twenty years in water and environmental issues, first as a postdoctoral scientist at the University of Minnesota in the United States and later at the Swiss Federal Institute of Aquatic Science and Technology (Eawag). From 1988 to 1992, he headed the water campaigns unit at Greenpeace Germany. Since 1992, he has been working as a consultant and advisor in the development of international environmental policies, and cooperates closely with universities, government agencies, municipalities, NGOs, and businesses.

Lars Müller
Lars Müller was born in Oslo. After his training as a graphic designer and student years spent in the United States and Holland, he founded his Atelier for Visual Communication and Design in 1982 in Baden. He has been a partner since 1996 of Integral Concept, an international affiliation for cross-disciplinary expertise, represented in Paris, Milan, Zurich, Berlin, and Montreal. Since 1983 he has published books on typography, design, art, photography, and architecture. Publications on social issues like human rights and ecology have supplemented the publishing house's catalogue since 2004. Since 1985 he has taught at various academic institutes in Switzerland and Europe, and since 2009 he has been a visiting lecturer at the Harvard University Graduate School of Design.

Christian Rentsch
Christian Rentsch, born in Zurich, pursued German studies, sociology, and musicology at the universities in Zurich and Berlin. During a career spanning more than thirty years, he worked as a freelance journalist, editor, and head of the culture section at the *Tages-Anzeiger* newspaper in Zurich. He is co-author and co-editor, together with Klaus Lanz, Lars Müller, and René Schwarzenbach, of *Who Owns the Water?,* published by Lars Müller Publishers.

René Schwarzenbach
René Schwarzenbach, Emeritus Professor of Environmental Chemistry at the Swiss Federal Institute of Technology (ETHZ), is currently Associate Vice President for Sustainability at ETHZ and head of the Competence Center Environment and Sustainability (CCES) of the ETH Domain. From 2004 to 2010, he was chair of the department of environmental sciences at ETHZ, and from 2006 until 2010, president of Division IV of the Swiss National Science Foundation. He received his diploma (1970) and his doctorate (1973) from the department of chemistry at ETHZ. After a postdoctoral fellowship at the Woods Hole Oceanographic Institute, he joined the Swiss Federal Institute of Aquatic Science and Technology (Eawag) in 1977, where he headed the department for multidisciplinary limnological research for several years and served on the board of directors until April 2005. In 1992, he was awarded the Körber Prize together with four colleagues from Germany, France, and Switzerland, and in 2006 he became the first non-American to receive the Award for Creative Advances in Environmental Science and Technology from the American Chemical Society. His textbook *Environmental Organic Chemistry* (1993, 2003) that he co-authored with two colleagues, has established itself as the standard work in the field.

ACKNOWLEDGMENTS

Many people have been committed to the creation of this book. They all deserve our special thanks.

We are very grateful to the members of the Sounding Board for their scientific support: Stefan Brönnimann (University of Bern), Andreas Fischlin (ETHZ), Nicolas Gruber (ETHZ), Gertrude Hirsch Hadorn (ETHZ), Volker Hoffmann (ETHZ), Reto Knutti (ETHZ), Christian Pohl (ETHZ), and Helmut Weissert (ETHZ).

We are also very grateful to all the professionals who contributed suggestions and critical comments on the soundness and comprehensibility of diverse topics and image sequences: Nathalie Baumgartner (ETHZ), Konstantinos Boulouchos (ETHZ), David Bresch (Swiss Re), Joan Davis, Dominik Fleitmann (University of Bern), Jean-Christoph Fueg (BFE), Jürg Fuhrer (Agroscope), Lino Guzzella (ETHZ), Stefan Hirschberg (PSI), Reinhard Madlener (RWTH Aachen), José Romero (BAFU), Roland Schertenleib, Karl Schuler (SRK), Stefan Schwager (BAFU), Thomas Spencer (FIIA), Johannes Staehelin (ETHZ), Michael Sturm, Max Ursin (KWO), Lasse Wallquist (ETHZ), Alexander Wokaun (ETHZ, PSI), Rolf Wüstenhagen (HSG), and Rainer Zah (Empa).

Florian Blumer and Florian Leu were very helpful in compiling information on extreme weather events in 2010 and drafting captions for the photos.

We also thank the team at Lars Müller's studio and publishing house. Astrid Lanz took over the copy-editing. Claudia Klein and Nadine Unterharrer viewed thousands of pictures from countless sources and assembled them in cohesive photo sequences. Linda Malzacher diligently persevered in identifying the ownership of pictures, and attended to copyrights in great detail. Nadine Unterharrer designed the book, and Esther Butterworth shaped its contents into printable form.

Finally, we thank Joanne Runkel and Alexis Conklin for their meticulous English translation and Jonathan Fox for his professional copyediting of this work.

The publication of this book would not have been possible without the trust and generous support of the following foundations:

AVINA STIFTUNG

HAMASIL STIFTUNG

GEORG UND BERTHA
SCHWYZER-WINIKER
STIFTUNG

FOR CLIMATE'S SAKE!
A VISUAL READER OF CLIMATE CHANGE

Edited by René Schwarzenbach, Lars Müller, Christian Rentsch, Klaus Lanz

Idea and concept: Christian Rentsch and René Schwarzenbach
Picture research and selection: Claudia Klein and Nadine Unterharrer
Copyrights: Linda Malzacher
Art direction: Lars Müller
Layout: Integral Lars Müller/Lars Müller and Nadine Unterharrer
Graphic artwork: The World as Flatland, Amsterdam, Netherlands
Production: Integral Lars Müller, Esther Butterworth
Translation: Joanne Runkel and Alexis Conklin
Copyediting: Jonathan Fox (English), Astrid Lanz (Original German)
Lithography: Photolitho AG, Gossau, Switzerland
Printing and binding: Kösel GmbH, Altusried-Krugzell, Germany

MIX
Papier aus verantwortungsvollen Quellen
FSC® C013736

Cover photo: Brown coal power plant in Cottbus, Germany. *Pawel Kopczynski/Reuters*

Lars Müller Publishers
Zürich, Switzerland
www.lars-mueller-publishers.com

ISBN 978-3-03778-245-3

Printed in Germany

Originally published in German, 2011.
The English translation was made possible with the support of the Swiss Agency for Development and Cooperation (SDC).
The views and opinions expressed in this book are those of the authors and do not necessarily reflect the official policy or position of the SDC.

Schweizerische Eidgenossenschaft
Confédération suisse
Confederazione Svizzera
Confederaziun svizra

Swiss Agency for Development and Cooperation SDC

Published with the support of

Zurich Insurance Company Ldt, Switzerland